Ageing and Degradation of Polymer Nanomaterials

Related Titles

Biopolymers Based Advanced Materials
ISBN: 978-0-6482205-4-1 (e-book)
ISBN: 978-0-6482205-5-8 (hardcover)

Functional Polymer Blends and Nanocomposites
ISBN: 978-0-6482205-6-5 (e-book)
ISBN: 978-0-6482205-7-2 (hardcover)

Functional Nanomaterials and Nanotechnologies: Applications for Energy & Environment
ISBN: 978-0-6482205-2-7 (e-book)
ISBN: 978-0-6482205-3-4 (softcover)

Advances in Polymer Technology: Material Development, Properties and Performance Evaluation
ISBN: 978-1-925823-00-4 (e-book)
ISBN: 978-1-925823-01-1 (hardcover)

Polymer Nanomaterials for Specialty Applications
ISBN: 978-1-925823-03-5 (e-book)
ISBN: 978-1-925823-04-2 (hardcover)

Advanced Materials
ISBN: 978-1-925823-05-9 (e-book)
ISBN: 978-1-925823-06-6 (hardcover)

Biofuels
ISBN: 978-1-925823-12-7 (e-book)
ISBN: 978-1-925823-13-4 (hardcover)

Liquid Crystalline Polymers
ISBN: 978-1-925823-16-5 (e-book)
ISBN: 978-1-925823-17-2 (hardcover)

Polymer Nanocomposites: Emerging Applications
ISBN: 978-1-925823-14-1 (e-book)
ISBN: 978-1-925823-15-8 (hardcover)

Ageing and Degradation of Polymer Nanomaterials

Dr. Vikas Mittal
Editor and Lead Author

CWP

Central West Publishing

NATIONAL
LIBRARY
OF AUSTRALIA

A catalogue record for this book is available from the National Library of Australia

ISBN (print): 978-1-925823-65-3
ISBN (ebook): 978-1-925823-67-7

Contents

Hyemin Lee, Zhou He, Yuting Li and Vikas Mittal

**6.   Effect of Hygrothermal Ageing on Mechanical,   113
      Thermal, Structural and Morphological Properties
      of Polypropylene and its Blends with Different
      Copolymers**
Anish Varghese and Vikas Mittal

**7.   Hygrothermal Ageing of Isotactic Polypropylene   143
      and Impact Polypropylene Copolymers: A
      Comparison of Mechanical, Thermal, Structural
      and Morphological Properties**
Anish Varghese and Vikas Mittal

Preface

A large variety of degradation environments have varying degrees of effect on the properties and performance of polymeric materials. In order to sustain the usefulness of the polymeric materials over longer periods of time, it is of vital importance to understand the various ageing and degradation phenomena taking place when the materials are subjected to different degradation environments. In this respect, the current books presents deep insights into the ageing and degradation processes of various commercially important polymeric nanomaterials.

Chapter 1 reviews the UV accelerated ageing and its influence on the performance of polymers and their blends as well as composites under various conditions of UV irradiance, temperature, moisture and time. The photo-degradation of diverse polymer coatings (e.g. acrylic, epoxy, polyurethane, silicone and polyester coatings) has been discussed in detail in Chapter 2. In Chapter 3, the biodegradation analysis of the bio-nanocomposites utilizing a blend of poly(butylene succinate) and chitosan with silica, silicate and graphene reinforcements has been performed. The biodegradability of the bio-nanocomposites was studied using the soil burial test under natural conditions followed by extensive thermal and morphological characterization. Chapter 4 presents a detailed explanation of various polyethylene degradation phenomena. The essential focus of the study is also to assess the corresponding stabilization needed to counter the various degradation processes. In Chapter 5, the degradation behavior of epoxy is briefly studied by focusing on the mechanism of induced damage. Stabilization of epoxy with additives to overcome the degradation phenomena is also briefly introduced together. In Chapter 6, the effect of hygrothermally induced ageing on the properties of neat polypropylene and its blends with ethylene vinyl acetate with maleic anhydride grafting, ethylene butyl acrylate and ethylene alpha olefin was studied. Simulated hygrothermal ageing conditions were created by immersing the polymer samples in distilled water and synthetic saline water. Further, in Chapter 7, hygrothermal ageing analysis of isotactic polypropylene and two different impact polypropylene copolymers was also carried out using different ageing conditions. In Chapter 8, the photo-degradation of isotactic polypropylene and impact polypropylene copolymers was studied in continuous UV weathering conditions. For this, injection molded specimens were exposed to a fluorescent light source for different time periods together with other accelerating factors.

The book would not have been successfully accomplished without the support of chapter contributors. The book is dedicated to my family for unswerving support, constant motivation as well as constructive suggestions for improvement.

Vikas MITTAL

1

UV Degradation and Weathering of Polymers, Blends and Composites

1.1 Introduction

Excellent property profiles and processability enable the polymeric materials to replace other conventional materials in application areas such as aerospace, automotives, marine infrastructure, building and construction, etc. In the case of outdoor applications, the overall performance of the polymeric materials greatly depends on their behavior towards various environmental factors including ultraviolet (UV) light, variations in temperature and moisture, pollutants, presence of oxygen and ozone, biological attack, mechanical stresses, etc. [1,2]. Lifetime predictions of polymers provide an idea about their durability, maintenance and replacement and are necessary for all kinds of outdoor applications. Commonly, the weatherability of materials can be anticipated using laboratory accelerated weathering testing with the help of artificial light sources such as xenon long arc, fluorescent UV, metal halide or carbon arc lamps. The molecular impact of accelerated weathering on the microscopic and macroscopic properties of polymers can be used to predict their weather resistance in natural use conditions [3,4].

Photo-degradation of polymers is an irreversible process in which the UV light decays their properties in the presence of other accelerating factors such as temperature, humidity, air pollutants, etc. The extent of interaction of the polymeric materials with UV light is the key factor determining the rate of photo-degradation [5,6]. The influence of UV radiations generally leads to two structural changes in the polymeric materials: chain scission and crosslinking. Chain scission upon UV weathering can result in decreased molecular weight and increased crystallinity with the evolution of more carboxylic acid and vinyl groups, whereas crosslinking generates increased molecular weight without affecting the crystallinity of the

Anish Varghese and Vikas Mittal, The Petroleum Institute (part of Khalifa University of Science and Technology), Abu Dhabi, UAE
Current address: Bletchington, Wellington County, Australia

materials. Amorphous region of the polymers is considered as the source for chain scission while the imperfect crystalline regions are the sources of crosslinking during UV exposure [7].

Previous studies have showed that Norrish Type I and Norrish Type II reactions are the basic routes for chain scission at the time of photo-degradation. Norrish Type I reactions result in the formation of free radical sites responsible for photo-degradation via photochemical scission of aldehydes and ketones, whereas Norrish Type II reactions create carbonyl and vinyl groups responsible for the formation of excited polymer-oxygen complexes. These reactions reduce the density of chain entanglements of the amorphous part in the semi-crystalline polymeric materials. The scission of highly entangled molecular chains reduces the molecular weight, thereby, leading to a considerable decrease in the mechanical properties [8,9].

In addition, the photo-degradation of polymers can be explained with the help of physical and chemical changes occurring in the polymeric materials in contact with UV light. Common causes for the chemical changes include activity of residual functional groups; monomers and photo-initiators; reactions caused by oxidation; hydrolysis; heat and UV radiations; presence of free radical occlusions; photo-cleavage products and peroxide putrefaction. Important sources of physical changes are inadequate interfacial adhesion; differential cure profile; curing generated volume contraction (internal stress); heat; mechanical forces (external stress and vibration); moisture and oxygen migration, etc. [10].

This chapter reviews the UV accelerated ageing and its influence on the performance of polymers and their blends as well as composites under various conditions of UV irradiance, temperature, moisture and time.

1.2 UV Degradation of Polymeric Materials

1.2.1 UV Degradation of Polypropylene (PP) based Materials

PP is one of the most versatile and cost effective thermoplastic polymeric materials with applications in packaging, fibers, automotive industries, non-durable goods, etc. Many studies have been carried out in order to understand the durability of PP based materials in outdoor applications (UV ageing) and also to assess their maintenance and replacement. Rabello and White [11] reported the photooxidative behavior of isotactic PP (iPP) using UV radiations of wavelength 290-320 nm at 30 °C for a duration of up to 48 weeks. Fourier transform infrared spectroscopy (FTIR) revealed the presence of degradation products such as carbonyls and hydroperoxides at the corresponding wavenumber ranges of 1700-1800 cm^{-1} and 3300-3600 cm^{-1}. Fractional crystallinity had a major role on the photodegradation of the polymer, and the least degradation was observed for the samples with higher crystallinity, however, significant effect of chemical degradation appeared only after the exposure period of 12 weeks. The strong dependency of the mechanical properties of the polymer on both chemical degradation and physical structure (spherulite size) was observed. Injection molded PP exhibited higher mechanical properties than the compression molded one owing to the larger spherulite size and absence of surface cracks. Most of the PP samples exhibited a tendency for partial recovery of the tensile performance after an exposure period of 6-9 weeks. Surface cracks on the tensile fractured surfaces were analyzed using scanning electron microscopy (SEM), and the presence of circular arc in the case of injection molded samples was observed, whereas randomly distributed cracks were noted for the compression molded samples. Differential scanning calorimetry (DSC) of the PP samples showed a decrease in the peak melting temperature (T_m) with increase in exposure time. Also, an increase in crystallinity was observed with increase in exposure time, which was attributed to the molecular scission of the highly entangled amorphous part and consequent rearrangement.

In another study, Yakimets *et al.* [12] examined the effect of photo-oxidation cracks on the mechanical properties of extruded iPP samples at 42 °C for a duration of up to 7 weeks. Crack formation was observed after 5 days of exposure. A decrease in the Young's modulus with crack growth up to 5 weeks was observed, followed by stabilization. The variations in the Young's modulus were divided into four periods: incubation period (3-5 days), decrease in modulus by cracks (5 days - 3 weeks), decrease in modulus by cracks and chemi-crystallization (3-5 weeks) and stabilization period (5-7 weeks). Also, an increased crystallinity was observed during 3-5 weeks duration and the decrease in the Young's modulus during this time was attributed to the formation of small independent crystallites by chemi-crystallization. Depth profiles of UV degraded PP were also reported by Nagai *et al.* [13], and the growth of degradation from the surface to the inner sections was observed in the range of micrometers only. This was attributed to the self-barrier effect of the degraded species through the absorption of UV light.

Shyichuk *et al.* [14] investigated the extent of chain scission and crosslinking of UV aged PP samples, along with the effect of tensile stress during ageing. Computer simulations of the molecular weight distributions (MWDs) from gel permeation chromatography (GPC) showed the dominating tendency of chain scission over crosslinking at the surface and inner bulk of the UV aged samples. Moreover, a strong influence of tensile stress on the chain scission of PP was observed during UV ageing. In another study, Shyichuk *et al.* [15] evaluated the extent of photo-degradation of stabilized and unstabilized form of two PP grades toughened with ethylene-propylene random copolymers.

The consequences of UV irradiation (λ<290 nm) and temperature (45 °C) on the structural transformations of syndiotactic PP (sPP) were also reported [16]. Before the analysis, the samples were crystallized from the melt and sPP with helical confirmations was used. For the exposure time of first 170 h, the crystallinity increased due to the enlarged crystalline domain size in the perpendicular direction of the (200) plane and the reduced trans-planar domains [16]. The role of temperature on the UV degradation of different grades of PP (homopolymer, random copolymer and impact copolymer) was also investigated by Tochacek and Vratnickova [17]. The accelerated ageing of different PP samples at various temperatures of 40, 50, 60 and 70 °C was compared with the natural outdoor aged samples based on the total UV radiation energy (TUVR) needed for the degradation. Among the different PP samples, the homopolymer exhibited higher UV stability, as compared to the impact copolymer. In another study, Francois-Heude *et al.* [18] measured the energy absorption of photosensitive species, responsible for creating radical sites and subsequently the commencement of deterioration in the presence of UV light, with the help of quantum theory. The authors also utilized two analytical models (Schwarzschild's law and kinetic model) to analyze the combined influence of UV light and temperature on the degradation induction time [18].

A composite material containing a hydrophobic thermoplastic as matrix (e.g. PP) and hydrophilic wood flour particles or fibers as reinforcement is termed as a wood-plastic composite (WPC). The advantages of such composites include enhanced environmental performance, low cost, lower density, reduced hazards, special aesthetic appearance, easy processability and recyclability. Due to the increasing demand of WPCs for the exterior structural and semi-structural applications (including decking, fencing, window framing, roof tiles, etc.), concerns have been raised about their long term performance and durability under the influence of UV light [19,20]. In this respect, PP/wood composites have received more research attention owing to their commercial relevance. Fabiyi *et al.* [9] studied the color fading and chemical changes of PP/wood fibers composites in the presence of UV light at 70 °C with a wavelength of 340 nm and compared these with the HDPE based WPC. Increase in the surface lightening, carbonyl concentration and wood loss were observed on increasing the exposure time, and the extent was higher in the PP based WPC as compared to the corresponding HDPE based WPC. The oxidation and consequent degradation of the lignin content in the wood fibers led to color fading and composite degradation upon UV exposure. In another study, La Mantia and Morreale [21] reported the influence of accelerated weathering on the mechanical properties of PP/wood flour composites. The WPCs were subjected to ageing cycle of 8 h irradiation at 55 °C and 4 h condensation at 35 °C. The mechanical properties of the WPCs were observed to be stable at shorter exposure time due to the stabilizing effect of wood flour, which was further confirmed by the molecular weight determination. Furthermore, a considerable UV degradation of WPCs was observed for longer UV irradiance time owing to the formation of chromophoric groups due to the deterioration of wood.

In order to examine the combined effect of UV light and moisture on the properties of WPCs containing bleached and unbleached kraft wood fibers, Beg and Pickering [22] studied the UV weathering of WPCs at 50 °C using an ageing cycle of 1 h UV irradiation (λ=340 nm) followed by 1 min water spray and 2 h condensation for a period of up to 1000 h exposure time. WPCs based on bleached fibers exhibited greater tensile strength, failure strain, impact strength and crystallinity when compared to the composites composed of unbleached fibers. This was attributed to the elimination of the unwanted non-cellulosic phase in the fibers and the subsequent improvement in the interfacial interactions between PP and bleached kraft wood fibers. Prolonged ageing conditions led to a reduction in the tensile strength, modulus and thermal resistance of both unbleached and bleached kraft wood fiber based composites, owing to the chain scission of PP and corresponding decrease in the polymer-fiber interactions. The authors recommended the use of stabilizer in the composites for achieving long life performance.

Butylina *et al.* [23] reported the influence of natural exterior weathering on the properties of PP/wood composites with and without pigments. The composites with dark pigments (brown and green) exhibited comparatively good discoloration resistance as compared to the composites containing other pigments. WPCs containing a higher fraction of PP presented a lower degree of surface crack generation and resultant degradation. Moreover, a reduction in the Charpy impact strength was observed due to the higher moisture absorption of the weathered samples.

In another study, Peng *et al.* [24] also reported the effect of accelerated weathering on the surface and mechanical behavior of PP/wood composites based on wood flour, lignin and cellulose for up to 960 h. The role of lignin content on the color fading of WPCs was confirmed using the colorimetric analysis. Photo-stabilization and antioxidation effects of lignin in the composites were underlined by the lower reduction rate of flexural strength and modulus, lesser extent of cracks and comparatively good surface hydrophobicity after ageing. The composites based on cellulose were stable towards the color changes during weathering, but had considerable reduction in the flexural properties. Atomic force microscopy (AFM) analysis of the composites revealed an increase in the surface roughness after UV weathering and the lignin containing composites showed a lower extent of reduction in the surface smoothness [25]. An increase in the crystallinity of PP in all composites was reported because of the chain scission and consequent re-crystallization of PP molecular chains in the composites during UV exposure.

Fabiyi and McDonald [26] examined the photo-degradation effect on MWD and crystallinity of PP/pine wood fibers composites using both natural and artificial ageing. Increased amount of surface cracks and pits was observed due to the higher extent of decomposition of wood fibers in the composites (60% wood and 40% PP) during weathering. Also, the wood decomposition left more PP domains on the degraded surface, noticed using pyrolysis gas chromatography-mass spectrometry. The chain scission of PP and the resultant highly ordered arrangements of the polymer chains in the composites increased the polymer crystallinity after weathering. An inverse relation of number and weight average molecular weight (M_n and M_w) with increase in UV exposure time was reported with the help of GPC. Furthermore, an increasing tendency of polydispersity was observed, which indicated the presence of secondary crosslinking process upon weathering.

The influence of UV ageing on the properties of wood/PP composites was evaluated by Kallakas *et al.* [20]. The composites containing smaller wood fractions exhibited good mechanical properties, especially flexural properties, owing to their more homogeneous structure. Further enhancement in the flexural properties was observed with coupling agent, 3-aminopropylytriethoxysilane (APTES), caused by the improved interactions between PP and wood flour through hydrogen bonding. The absorption of both UV radiation and water molecules during the weathering process reduced the flexural properties of WPCs, whereas an increase in the impact strength was observed. The improvement in the impact strength was attributed to the weaker interfacial bonding and the corresponding formation of various cracks at the time of sudden force. In another study, Butylina *et al.* [19] reported the UV accelerated ageing behavior of WPCs consisting of three different mineral fillers such as calcium carbonate, wollastonite and talc upon exposure for up to 2000 h. The results of the color measurements showed greater color fading tendency of the mineral filler containing composites than pure WPC. The surface degradation of the PP layers enveloping the wood fibers was observed in pure WPC. At the same time, the surface deterioration of PP in the mineral containing WPCs was lower when compared to pure WPC. FTIR studies revealed a reduction in the lignin index with increase in irradiance exposure time for all WPCs, and the degree of decrease was lower in the case of mineral fillers containing WPCs. Also, the carbonyl index of all composites was observed to increase with increase in irradiance period. Furthermore, WPC containing talc exhibited greater retainability of the Charpy impact strength as compared to other composites because of the compatibility between talc and PP. Darie *et al.* [27] studied the properties of PP/eucalyptus wood fibers/poly(lactic acid) (PLA) ternary composites before and after exposure to UV light of wavelength ranging from 200-700 nm at 30 °C and 60% humidity for 600 h. Toluene-2,4-diisocyanate (TDI) was used for the chemical modification of wood fibers to enhance their compatibility with the polymer matrices. The improved interfacial interactions of the constituents, thermal stability and slightly enhanced mechanical properties were reported for the TDI modified wood fibers based composites. The composite containing 70% PP, 15% TDI modified wood fibers and 15% PLA exhibited good properties after UV ageing.

Only few studies have been reported regarding the photodegradation of PP composites reinforced with natural fibers other than wood [28,29]. Joseph *et al.* [30] reported the UV weathering effect on the tensile properties of the PP composites reinforced with sisal fibers. The composites reinforced with unmodified and chemically modified sisal fibers were subjected to UV irradiation for up to 12 weeks. A decrease in the tensile properties was observed with an increase in irradiation period for all fiber loadings. This was attributed to the chain scission of PP owing to photo-oxidation. Also, a substantial decrease in the tensile properties was reported for modified fibers based PP composites. The recovery of the tensile properties was improved with fiber loading after UV ageing. Moreover, SEM images revealed the ability of fibers towards crack propagation control in the composites. Photo-degradation behavior of the un-compatibilized and compatibilized date palm fibers/PP composites was also evaluated by Abu-Sharkh and Hamid [28]. The authors observed appreciable stability of the composites under UV weathering in both natural and accelerated weathering conditions. The presence of lignocellulosic fibers imparted higher UV resistance to the composites. The reduction in the melting temperature of the composites was observed to be very low as compared to pure PP after accelerated UV exposure. Due to the lower stability of the compatibilizer, maleated PP exhibited faster photo-degradation of the compatibilized composites as compared to the un-compatibilized composites.

Gadioli *et al.* [31] evaluated the effect of lignin content in the PP/cellulose composites against UV weathering. Both bleached and semi-bleached fibers reinforced PP composites were used for this purpose. Semi-bleached cellulose fibers reinforced PP composites exhibited excellent weathering stability for all lignin contents compared to the bleached fibers based composites. The presence of hindered phenol in the lignin content of the semi-bleached fibers offered outstanding photo-stability to the composites. Moreover, the presence of strong adhesion between the fibers and PP matrix after UV ageing was revealed using SEM.

Long fibers reinforced thermoplastics (LFTs), composed of fibers (glass, carbon or aramid fibers) with aspect ratios in the range of 200-3000, are an important class of fibers reinforced composites. The high stiffness and affordable processing cost of LFTs drive their overall growth [32,33]. Goel *et al.* [34] examined the influence of UV irradiation on the microstructure and mechanical properties of PP/E-glass LFT. The percentage crystallinity of the damaged LFT surface was increased with increasing period of UV exposure, which was caused by the chemi-crystallization of PP in the LFT. An increase in the Young's modulus at the degraded surfaces of both LFT and PP with increase in UV exposure period was reported with the help of nanoindentation. The increase in the modulus in LFT was faster as compared to PP because of the presence of added chromophores during the sizing of glass fibers. At the same time, the dynamic Young's modulus of LFT was reduced due to the development of surface cracks caused by the volume change.

In another study, Chevali *et al.* [35] studied the UV accelerated degradation effect on the creep behavior of PP based LFT. Increasing the UV exposure time formed cracks on the surface of LFT, along with color fading. UV degradation induction period of PP LFT was observed to be in the range of 300-400 h. Flexural creep testing of the PP LFT samples revealed an increase in creep compliance with exposure duration. For instance, about 33% increase in creep compliance was observed with increase in exposure time from 100 h to 400 h for the analysis at 1000 s with a constant load of 64 N.

Rabello *et al.* [36] investigated the accelerated photo-degradation behavior of PP/glass fibers composites containing weld lines and compared the results with PP composites containing talc and calcium carbonate. Significant influence of filler aspect ratio on the debility of weld regions was noted, and the fillers with higher aspect ratio led to an increased debility of weld regions. Mechanical characterizations after UV exposure indicated the worsening of mechanical properties, and it was greater in the case of PP and PP/glass fibers composites. The decay in the mechanical properties of the PP/glass fibers composites resulted from the loss of adhesion between PP and glass fibers during UV ageing.

Among other composites, the UV weathering performance of polymer nanocomposites is important to analyze so as to widen their engineering applications in external environmental conditions [37,38]. In this respect, photo-oxidation of PP/clay and PP-*graft*-maleic anhydride (PP-g-MA)/clay nanocomposites containing 5 wt% of filler was evaluated by Tidjani and Wilkie [39] using UV irradiation at $\lambda<300$ nm at 50 °C. The unsuitability of clay towards the photo-stability of PP was observed, and the disappearance of photo-degradation induction period of nanocomposites confirmed their faster photo-degradation than PP. Mailhot *et al.* [40] also reported similar chemical changes and photo-degradation behavior of PP-g-MA compatibilized PP/organo-modified montmorillonite nanocomposites.

Bocchini *et al.* [38] studied the photo-oxidation of PP nanocomposites comprising of unmodified and organically modified nano-boehmite at 60 °C with λ>300 nm. The presence of unmodified boehmite in the PP matrix reduced its photo-oxidation induction period, which was attributed to the adsorption of phenolic antioxidant present in PP on the hydrophilic polar boehmite surface. This was further confirmed using the photo-degradation of solvent extracted PP/boehmite nanocomposites. In order to avoid the adsorption of antioxidants on the boehmite surface and thereby to achieve UV stability, organic modification of boehmite using *p*-toluenesulphonate (PTS) or *p*-alkyl benzylsulphonates (PABS) was also carried out. The authors observed that the chemical modification of boehmite using PTS was effective in retaining the performance of antioxidant, while PABS increased the oxidation rate due to the creation of radical sites. In another study, Bocchini *et al.* [41] reported similar oxidation induction time (OIT) reduction behavior for PP-g-MA compatibilized PP/hydrotalcite nanocomposites as compared to pristine PP. Also, the photo-stability recovery of the solvent extracted PP based nanocomposites was reported. Furthermore, the authors observed the photo-oxidation acceleration effect of the transition metal present as impurity in nano-hydrotalcite.

Alongi *et al.* [42] investigated the role of β-cyclodextrin nano-sponges (NS) on the UV accelerated ageing of PP in combination with UV stabilizers such as 2-hydroxy-4(octyloxy)-benzophenone (HOBP) and triphenyl phosphite (TPP). PP/NS nanocomposites presented an increase in absorbance at hydroxyl (2723 cm^{-1}) and carbonyl (1717 cm^{-1}) infrared regions. It was observed that the incorporation of NS alone in PP reduced its OIT from 12 h to 8 h during UV ageing. This proved the unsuitability of NS against photo-oxidation. The incorporation of 1 wt% of HOBP increased the OIT of PP remarkably to 38 h and decreased the photo-oxidation rate to 1.1 x 10^{-2} h^{-1} from 5.2 x 10^{-2} h^{-1}, whereas 1 wt% of TPP reduced the OIT to 28 h without affecting the oxidation rate.

1.2.2 UV Degradation of Polyethylene (PE) based Materials

Polyethylene is a popular cost effective commodity thermoplastic polymeric material with excellent properties such as light weight, high strength, water resistance and stability. These properties result in its application in many areas, especially packaging, electrical insulation and agriculture [43]. In general, various types of PE are available, including high density PE (HDPE), low density PE (LDPE), linear low density PE (LLDPE), ultra-high molecular weight PE (UHMWPE) and metallocene PE.

Dehbi *et al.* [44] studied the effect of UV ageing at 40 °C and 50 °C on the physical, mechanical and structural properties of the LDPE tri-layer films for duration of up to 8 months. Proportional increase in the surface free energy with temperature was noticed, and the increase became more noticeable in combination of UV irradiation and temperature. The breaking of PE molecular chains during UV exposure correspondingly decreased the yield stress and elongation at break of the films. FTIR spectra of the aged PE samples exhibited the presence of the peak at 900-1040 cm^{-1} after 500 h exposure due to the vibration of CH_2-O groups or double bonds. Also, the intensity of this peak was observed to increase with ageing time. Dehbi *et al.* [45] further confirmed the decreased mechanical performance (except elastic modulus) of the PE films under natural and artificial ageing conditions. The increase in the modulus of elasticity was considered as a deterioration resulting due to the ductility loss of PE.

In another study, Dehbi *et al.* [46] also reported the effect of climatic conditions (UV radiations, temperature and humidity) on the thermo-mechanical performance and the durability of the PE films. The heat of fusion and degree of crystallinity of the PE films were increased with ageing, whereas the melting temperature had no remarkable variation. The decrease in the mechanical performance of the PE films was observed to be optimal at combined conditions of 50 °C and UV irradiation. The authors concluded that the effect of UV weathering diminished the performance as well as durability of the PE films. Imane *et al.* [47] also reported an increase in the crystallinity of the PE films with UV ageing. It was observed that the crosslinking process at the initial stages and chain scissions at the following stages of ageing, along with the resultant chemicrystallization of the ruptured PE chains, increased the crystallinity of the UV aged PE films.

Mohamed *et al.* [48] investigated the nanoscale mechanical properties of UHMWPE after UV ageing. The generation of crosslinking under the influence of UV exposure was observed, which correspondingly resulted in increased nano-hardness as well as modulus of elasticity, along with decreased viscoelastic behavior.

Stark and Matuana [49] examined the influence of weathering on the surface behavior of injection molded, extruded and planed HDPE/wood flour (pine) composites using UV radiations of λ=300-

400 nm for up to 3000 h. A weathering cycle of 108 min UV irradiance followed by 12 min water spray was used. The deterioration of HDPE and swelling of wood fibers during weathering resulted in the surface cracking of the composites. The use of stabilized HDPE correspondingly reduced the surface cracking of the composites. The surface of the injection molded samples was observed to be more resistant towards deterioration due to a lower degree of surface voids. The decrease in FTIR peak intensities at 3318 and 1023 cm^{-1} also showed the deterioration of wood on the composite surface. Also, a decrease in wood index was observed for all samples. Extruded and planed composites exhibited higher contact angle than the injection molded samples, which was attributed to the lubricant used at the time of extrusion. In another study, Stark *et al.* [7] reported a higher degree of surface degradation for HDPE/wood flour composites than pure HDPE after a weathering period of 2000 h. This was observed from the greater oxidized-to-unoxidized carbon ratio of 80% for the composites as compared to 5% for pure HDPE. The presence of wood in the composites inhibited the normal cross-linking of the HDPE chains during the initial stages of ageing, thus, resulting in an earlier chain scission of the HDPE chains and analogous enhancement in crystallinity. Lundin *et al.* [50] also reported a lower retention degree of the mechanical properties (flexural strength and strength) for the PE/wood composites than pure HDPE during weathering.

The improvement in the weathering resistance of the HDPE/wood composites by the incorporation of UV absorber (hydroxyl phenyl benzotriazole) and pigment (zinc ferrite in a carrier wax) was reported by Stark and Matuana [51]. As compared with pure HDPE, the addition of both UV absorber and pigment resulted in less color fading after 3000 h of weathering. Increased color stability was observed on increasing the concentration of pigment in the composite formulation, while an increase in the UV absorber concentration had no remarkable effect. Also, enhanced recovery in the mechanical properties of the WPCs was observed after incorporating UV absorber or pigment. Both flexural strength and elastic modulus had enhanced recovery on increasing the UV absorber content, whereas an increase in the pigment content improved the flexural strength only. The observed behavior resulted from the action of the carrier wax covering the pigment, which created hydrophobic surfaces and correspondingly prevented the moisture absorption. In another study, Fabiyi *et al.* [52] studied natural as well as accelerated weathering behavior of WPCs based on various modified wood fibers. The composites based on holocellulose fibers exhibited more wood fibers on the surface as well as higher color fading resistance after ageing whereas both pine and extractive-pine wood fibers increased the color change and photo-oxidation of the composites. Also, the crystallinity of the holocellulose based composites remained unaltered during UV exposure while the crystallinity of the pine based composites increased owing to the decreased M_w of PE. Fabiyi and McDonald [53] also reported substantial color stability for the PE/wood composites based on hybrid poplar and ponderosa fibers.

To improve the compatibility between PE and wood, Krehula *et al.* [54] made use of diisodecyl phthalate (DIDP) plasticizer. Changes in the surface chemistry and thermal stability after UV weathering were studied for pristine composites and composites with varying content of Tinuvin stabilizer. Good weathering resistance performance was observed for the composite containing 0.25 wt% of Tinuvin. The inclusion of 0.25 wt% of Tinuvin generated only 38% carbonyl index as compared to 114% for pristine composite, thus, indicating lower degree of degradation.

Kumanayaka *et al.* [55] studied the UV weathering of LDPE/PE-g-MA/clay nanocomposites at 60 °C. An increased UV degradation rate was observed after 2 weeks of UV exposure. The presence of Fe in the organo-modified clay as well as the break-up of alkyl ammonium ions of clay caused an increase in the UV degradation rate with increase in clay content. To study the role of organo-modifier amount as well as polymer matrix polarity on weathering, Botta *et al.* [56] performed the UV weathering of LDPE/EVA/clay nanocomposites composed of different organo-modifier concentration and PE/EVA blend ratio. Clays with higher amount of organo-modifier generated faster degradation rate in the nanocomposites as compared to the clays containing lower organo-modifier content.

The accelerated weathering of LLDPE/clay nanocomposites was also reported by Dintcheva *et al.* [57], using organo-modified and pristine montmorillonite clays with or without PP-g-MA compatibilizer. The samples were aged using a weathering cycle of 8 h UV irradiation at 55 °C and 4 h condensation at 35 °C. The layered silicate clays had no effect on the weathering behavior of PE. A decrease in the photo-stability was reported for PE-g-MA compatibilized PP/organo-montmorillonite nanocomposites as compared to the nanocomposites without compatibilizer. The authors opined that the factors such as adverse reactions of anhydride with clay as well as surfactant decomposition products, UV instability of anhydride and unfavorable catalytic influence of metallic ions in the clay decreased the photo-stability of the nanocomposites.

In another study, Grigoriadou *et al.* [58] examined the mechanical, thermal and surface properties of HDPE nanocomposites based on unmodified and modified montmorillonite (MMT), multi-walled carbon nanotubes (MWCNTs) and SiO_2 after accelerated UV weathering at 25 °C with a wavelength of 280 nm. HDPE/MWCNTs nanocomposites exhibited more stabilized tensile modulus and elongation at break, indicating that the UV stabilizing effect of MWCNTs in HDPE. FTIR spectra also supported the stabilizing effect of MWCNTs in HDPE, whereas the spectra of nanocomposites based on other nanomaterials revealed the formation of carbonyl and hydroxyl groups, along with a small amount of vinyl compounds. Also, SiO_2 was observed to result in faster UV degradation during ageing. Increase in crystallinity was noted for all nanocomposites, however, its magnitude was low for HDPE/MWCNTs nanocomposites owing to the photo-stabilizing effect of MWCNTs, which protected the HDPE chains from breaking. Dintcheva *et al.* [59] also reported the weathering behavior of the LDPE/MWCNTs nanocomposites. Higher amount of MWCNTs (1 and 2 wt%) in LDPE exhibited a lower degree of increase in the mechanical properties as well as decreased carbonyl formation at prolonged irradiation time, whereas opposite behavior was noticed for lower MWCNTs contents (0.1, 0.2 and 0.5 wt%).

Improved UV stability of HDPE was accomplished by Grigoriadou *et al.* [60], through the incorporation of varying amount of Cu-nanofibers. Increased stability for all nanocomposites was observed after UV irradiance (280 nm) for various days. The observed oxidation depth profiles were lower for Cu-nanofibers based HDPE composites because of reduced oxygen permeability, which resulted in improved photo-oxidation stability. Moreover, the authors underlined the UV degradation mechanism through the molecular chain scission as well as the creation of carbonyl, hydroxyl and unsaturated groups containing compounds, using FTIR analysis. Yang *et al.* [61] observed reduced stability of LDPE/ZnO nanocomposites upon UV irradiation owing to the enhanced generation of carboxyl groups as well as CO_2. The authors compared the observations with the LDPE composites with TiO_2 and observed a greater degree of carbonyl group formation for TiO_2 and greater CO_2 generation for ZnO.

1.2.3 UV Degradation of Epoxy based Materials

Epoxies are remarkable class of thermosetting polymers obtained from the condensation reaction of bisphenol-A (BPA) and epichlorohydrin (ECH). Crosslinked structures of epoxies are obtained by using hardeners such as aliphatic amines, aromatic amines, polyamines and anhydrides. The properties of epoxies can be tailored for various applications depending on the chemical nature of hardeners and curing conditions. Epoxy resins have outstanding properties such as high strength and modulus; good resistance to chemicals and moisture; effective resistance to creep, abrasion and corrosion; adhesion to other substrates; excellent fatigue resistance; good electrical resistance, etc. The excellent properties of epoxy resins drive their use in many applications including engineering and structural composites, protective coatings, high strength adhesives, etc. [62,63]. Weathering of epoxy resins is the only major factor limiting their outdoor applications. This is attributed to the presence of aromatic components in their structure, which absorb UV radiations in the range of 300 nm and, thus, undergo chain breaking and cross-linking, resulting in the degradation of the matrix [63,64].

Higher retention in the comprehensive mechanical properties of the alkali treated hemp fibers based epoxy composites on weathering was reported by Islam *et al.* [65], as compared to untreated hemp fibers based materials. This was attributed to the increased fiber-epoxy interactions after alkali treatment. Also, the mechanical properties of both composites upon ageing were higher than pristine epoxy resin. The mechanical properties of the samples decreased with exposure time owing to the loss of fiber-epoxy interactions caused by the swelling and corresponding crack development during moisture intake. The authors also observed an increase in the impact energy for both composites owing to the plasticizing effect of the entrapped water molecules in the composites, which entered in the composites through the porous structure formed after the decomposition of lignin [65]. In another study, Dixit [66] also observed a reduction in the mechanical properties (tensile and flexural strength) with UV exposure for the coir-epoxy infused particulate composites.

Yan *et al.* [67] also noticed a decrease in the mechanical properties of the epoxy/flax fabric composites with exposure time. About 30% and 35% decrease in the tensile strength and modulus after 1500 h exposure was observed respectively, whereas the decrease in the flexural strength and modulus was about 10%. Also, the reduction in the mechanical properties of composites was lower than pure epoxy. SEM analysis revealed a decrease in the epoxy-flax fabric bonding on weathering, which subsequently decreased the mechanical properties.

Removal of epoxy matrix from the surface of epoxy/glass-fiber composites was observed by Dubois *et al.* [68] under natural and accelerated weathering conditions. Kumar *et al.* [69] also observed an erosion of the surface epoxy matrix during the accelerated weathering of epoxy/carbon fiber (CF) composites. This simultaneously affected the mechanical performance of the composites. For example, about 29% reduction in the tensile strength was noted after 1000 h exposure due to the failed load transfer from the eroded epoxy matrix to CF. Singh *et al.* [70] reported further degradation of the epoxy layers below the weathered surface layer of the epoxy/CF composites on the application of *in-situ* mechanical loading during weathering.

The role of reinforcement characteristics on the photo-degradation of the epoxy composites was reported by Awaja and Pigram [71] for the composites reinforced individually with E-glass (E), 3D-glass (3D) and CF. The authors noted higher extent of oxidation on the exposed surface of the epoxy/3D composites as compared to other samples. Also, the presence of both chain scission and crosslinking in CF composites was noted during UV irradiance, while only chain scission was noted for the other two composites. In another study, Awaja *et al.* [72] confirmed the UV accelerated damage in the form of delamination in CF composites, spherical voids in 3D composites and a combination of crack, microcrack and delamination in E composites.

The effect of nanomaterials on the long term weathering performance of epoxy/glass fibers composites was also examined by Chang and Chow [73]. The authors observed good retention of the flexural properties for the epoxy/glass fibers/org-MMT ternary nanocomposites after an exposure time of 100 h at 50 °C. Furthermore, the fracture toughness of the nanocomposites decreased significantly upon weathering, which was caused by the sensitivity of the notch towards moisture and temperature. In another study, Zainuddin *et al.* [74] reported that the incorporation of 2 wt% nanoclay was beneficial in enhancing the mechanical properties of the epoxy/glass-fiber composites throughout the weathering conditions of UV irradiation as well as combined UV irradiation and condensation, as compared to pure and 1wt% nanoclay composites. This was attributed to the improved interfacial bonding between the components and resultant reduced delamination of the composite, which was confirmed using both optical and SEM micrographs. Moreover, the addition of nanoclay enhanced the stability of epoxy/glass-fiber composites towards color fading and weight change.

The effectiveness of the epoxy/organoclay nanocomposites towards moisture degradation during weathering caused by the combination of UV radiations and moisture was reported by Woo *et al.* [75], resulting from the outstanding moisture barrier performance of nanoclay. The authors also reported improved color stability for the organoclay nanocomposites. In another study, Woo *et al.* [76] observed good mechanical properties for the epoxy/organoclay nanocomposites on weathering. The observed retention in the flexural modulus was higher as compared to neat epoxy. The tensile strength of the nanocomposites showed an increasing tendency after an exposure of 250 h. The extent of micro-hardness and nanoindentation modulus increase at the weathered surfaces of the nanocomposites was lower as compared to neat epoxy. Mailhot *et al.* [77] also confirmed the significant recovery in mechanical properties for the epoxy/clay nanocomposites upon exposure to UV radiations of $\lambda > 300$ nm at 60 °C. Photo-oxidation stability was observed to be higher for the epoxy/clay nanocomposites as compared to the silica based composites. Recently, Shanmugam *et al.* [78] also observed excellent weathering resistance of oil fly ash (OFA) based epoxy composites. The observed mechanical, thermal and morphological behavior of 4 wt% OFA composite after weathering was outstanding when compared with 0.5 wt% commercial light stabilizer (Hostavin® N24) filled epoxy composites.

1.2.4 UV Degradation of Polyester Resin based Materials

Polyesters are macromolecules containing ester linkages in their backbone chain, formed by the reaction of dihydric alcohols and dicarboxylic acids. Generally, polyesters are classified as unsaturated (UP) and saturated polyesters. UP resin is a dominant thermosetting material owing to its enhanced mechanical performance, easy processability and low cost. These advantages of UP drive it use as the matrix material for engineering and structural composites. Preparation of UP includes prepolymer generation and crosslinking steps. Prepolymer is obtained from the reaction of unsaturated dibasic acid (maleic anhydride and fumaric acid) and saturated acid (phthalic acid and glycol), which is then dissolved in a vinyl monomer (styrene) to allow successful crosslinking in the presence of crosslinking agent (peroxide or hydroperoxide) [79,80].

Increased wettability of UP resin during accelerated weathering was reported by Jia *et al.* [81] due to the generation of hydroxyl and carbonyl groups. The weathering cycle included 4 h UV irradiance (313 nm) at 50 °C followed by 4 h condensation at 50 °C. Microscopy analysis revealed the crack formation with weathering time. The gloss exhibited an increase up to 12 days, followed by a decreasing tendency. Also, color stability was noted after the first stage of weathering. Gu *et al.* [82] investigated the UV stability of the UP/glass fibers composites for short irradiation period. Composite laminates consisting of 2, 3 and 5 layers were analyzed. Poor penetration of the UV radiations inside the composites was reported, with reactions observed only at the surface and in slight depths. UV radiations had no remarkable influence on the tensile strength of the 5 layer composites whereas the 3 layer sample exhibited degradation in tensile strength after 200 h exposure. In another study, Hongwang *et al.* [83] also observed decreased bending and tensile strength for 2 layer laminates of the UP/glass-fibers composites. This was attributed to the erosion of the PU matrix and corresponding debonding between PU and glass fibers. Increased stiffness of UP/20 wt% glass fibers composite was reported by Mouzakis *et al.* [84] after ageing up to 6 weeks following a cycle of 4 h UV radiation at 60 °C and 4 h condensation at 50 °C. The observed behavior resulted from the post-curing of UP in the presence of UV light and temperature.

Rodrigues *et al.* [85] examined the UV weathering of the UP/E-glass/curaua hybrid composites for 84 days, and the observed degradation extent was higher as compared to the UP/E-glass composites after ageing. It was attributed to the degradation of curaua fibers. In another study, Shahzad [86] reported about 3% weight loss for the PU/hemp fibers composites when exposed to UV irradiation alone for 1000 h, however, it was about 2% for weathering caused by the combined action of UV radiation as well as condensation, which was attributed to the dissociation of UP in the composite. Moreover, stabilized tensile properties of the composites were observed after an initial reduction at the beginning of the ageing process. The reported reduction in the tensile strength was about 30% for the composites aged under both conditions.

1.2.5 UV Degradation of Polyurethanes (PUs) based Materials

PUs are useful polymeric materials obtained from the polyaddition reaction of polyisocyanates and macro-polyols. These materials are available in thermoplastic, thermoset and elastomeric forms based on the nature of polyisocyanates and polyols as well as crosslinking density. The properties of PUs include excellent mechanical performance, good chemical and electrical resistance, processability, outstanding tear propagation resistance, etc., thus, making them useful for extensive commercial, technical and engineering applications [62,87]. However, high UV sensitivity and resulting structural changes of PUs during outdoor exposure reduce their durability owing to the degradation of the physical and mechanical characteristics. In contact with UV radiations, aromatic urethanes present in PUs undergo photo-oxidation by means of quinonoid pathway. This generates powerful chromophoric structure of quinone-imide, resulting in yellow coloration [88,89].

Rosu *et al.* [90] observed color changes during the photo-degradation of PU when exposed to UV radiations of 300 nm for 200 h. The degradation mechanism included the scission of urethane groups and corresponding oxidation of methylene groups present in between the aromatic rings. The increasing tendency of stress at 200% strain and tensile modulus of PU after an initial decrease at the beginning of UV ageing was observed by Boubakri *et al.* [91], which was attributed to the increased crosslink density due to prolonged ageing. Similar behavior was observed for the glass transition temperature (T_g) upon ageing. A decrease in the wear resistance was also noted throughout the exposure time. Saadat-Monfared and Mohseni [92] examined the role of nano-cerium oxide on the UV ageing of the PU films and observed an enhanced T_g with an increase in the nano-cerium oxide amount as well as weathering time. The observed effect might have resulted due to the reduced photo-degradation (oxidation and chain scission reactions) in the nanocomposites owing to the higher UV photo-absorption of nano-cerium oxide.

To improve the UV stability of PU elastomers, Zia *et al.* [93] proposed the use of chitin as chain extender, used along with 1,4-butane diol (BDO) extender as blend in different compositions (0, 25, 50, 75, 100 wt%). PU elastomer with a higher extent of chitin (i.e. 100 wt% as extender) presented greater photo-stability as compared to other matrices. In another study [94], good retention in the surface properties was observed for PU elastomers with higher chitin content upon ageing, due to the increased hydrophobicity on chitin addition [94]. This was further confirmed using PU elasto- mers extended with a combination of chitin, BDO and dimethylol propionic acid (DMPA) in various compositions [95].

1.2.6 UV Degradation of Polycarbonate (PC) based Materials

PC is an engineering thermoplastic polymeric material with car- bonate intergroup linkages in its molecular structure, produced by the interfacial polycondensation of bisphenol A and phosgene or the ester interchange reaction between bisphenol A and diphenyl car- bonate. PC finds applications in automotive, construction, electron- ics, appliances and lighting because of its superior properties such as low temperature impact strength, good stiffness, favorable heat and electrical resistance, ability to transmit light, excellent creep resistance, etc. [80,96]. The photo-Fries reaction and photo- oxidation upon long term weathering results in the deterioration of toughness and ductility, along with discoloration. Tjandraatmadja *et al.* [97] observed these changes in PC during weathering for up to 3000 h. A significant change in the refractive index of the PC films upon UV irradiation was observed by Migahed and Zidan [98]. Moreover, a decrease in the optical energy gap and an increase in Urbach energy were noticed. Perez *et al.* [99] examined the influ- ence of UV ageing on the mechanical and thermal properties of up to 10 time reprocessed PC. Thermal properties were observed to de- crease with increasing reprocessing cycles for both aged and unaged samples. The mechanical properties (tensile strength and modulus) were unaffected up to 7 reprocessing cycles before ageing, whereas a decreasing tendency was noticed for the aged samples after first reprocessing cycle. Sloan and Patterson [100] revealed the rapid yellowing of PC/layered silicate nanocomposites as compared to neat PC upon exposure to UV radiations of 340 nm. A correlation between the carbonyl breakage and discoloration of PC was also de- veloped.

1.2.7 UV Degradation of Vinyl Ester Resin (VER) based Materials

Vinyl ester resins (VER) are thermosetting polymers containing double bonds as the end groups and are prepared by the esterification of epoxy resin with α,β-unsaturated acids. These materials are commonly used as the matrix material for engineering and structural composites due to their properties such as outstanding strength, good thermal stability, longevity, corrosion resistance, etc. However, the influence of UV irradiation is observed to be detrimental on the outdoor performance of VER, which is related to the photo-oxidative or photolytic reactions on interaction with UV [101,102].

Signor *et al.* [103] noticed about 40% and 60% decrease in the corresponding strain to failure and toughness of VER after UV exposure of 4000 h, which was related to its transition from ductile to brittle nature during UV exposure. Moreover, a notable increase in the hardness and modulus was observed after an exposure of 1000 h due to the loss of ductility. The generation of surface defects and chemical changes with exposure time was noted using respective microscopy and spectroscopy analysis. The decrease in the surface gloss as well as T_g of the cured vinyl ester network (VEN) during UV ageing was demonstrated by Rosu *et al.* [102], which was attributed to the scission of VEN molecular chains upon UV irradiation (300-540 nm). Also, the generation of carbonyl, hydroxyl, peroxides and hydroperoxide based degradation products was observed. The generation of unstable hydroperoxides resulted in the formation of polyhydroxy ether as well as saturated ester structures, along with volatile CO and CO_2.

UV accelerated degradation behavior of a blend of VEN and ammonium lignosulfonate (ALS) modified lignin was studied by Rosu *et al.* [104], and the presence of ALS in the blend improved the photostability to VEN to a small extent. Korach *et al.* [105] reported the ageing behavior of VER/carbon fiber composites under various combinations of UV irradiation, salt spray, temperature and humidity. The authors observed that the combination of UV radiation and salt spray had more detrimental effect on the properties of the composites. Both flexural strength and energy release during tensile test decreased after UV weathering, whereas the bending modulus remained stable.

1.2.8 UV Degradation of Poly(vinyl alcohol) (PVA) based Materials

Poly(vinyl alcohol) (PVA) is a biocompatible water soluble polymer obtained from the hydrolysis or alcoholysis of polyvinyl acetate and possesses unique properties such as hydrophilicity, biodegradability, non-toxicity, etc. [106]. In PVA, the presence of acetic side groups, branching points, aldehyde end groups, tail-to-tail structures, etc., can induce photo-degradation in contact with UV radiations. Kaczmareck *et al.* [107] studied the UV accelerated ageing behavior of the PVA/MMT nanocomposites and observed non-uniform variation in the chemical and physical properties. GPC analysis revealed chain scission during UV exposure as a result of photo-oxidation, while a low extent of crosslinking was noted. Moreover, the addition of MMT reduced the amount of ketonic group generation during photo-oxidation of PVA.

Photo-stability of PVA was achieved by Kaczmarek and Podgorski [108] through the use of graphite oxide (GO). As a result of photo-oxidation, more functional groups such as carbonyl, hydroxyl and ether were observed on the surface of GO, which resulted in effective bonding with PVA. PVA composite with 5 wt% GO had good thermal stability after UV ageing, whereas a small decrease in the thermal properties was observed for the composites with lower GO content (<1 wt%). The increased thermal stability of the PVA/5 wt% GO composite was attributed to the photo-crosslinking at the time of UV exposure. The enhancement in the UV stability of PVA with GO incorporation was further confirmed by Moon *et al.* [109]. The PVA composites were generated with varying amount of oxyfluorinated GO (1, 3, 5 and 10 wt%) and crosslinked with glutaraldehyde. The decreasing tendency of the soluble gel content of the UV aged samples was observed on increasing the GO content. The observed behavior was attributed to the increased interactions between PVA and GO, which resulted from the fine dispersion of functionalized GO in the PVA matrix.

1.2.9 UV Degradation of Other Polymers and Copolymers

The effect of both natural and accelerated UV weathering on polyvinyl chloride (PVC)-wood composites was reported by Chaochanchaikul *et al.* [110]. The rate of deterioration was lower in natural ageing conditions for both PVC and PVC composites as compared to accelerated ageing. Also, the suitability of 2 wt% UV stabilizer, Tinuvin P, for the photo-stabilization of the samples was demonstrated. The presence of wood had a minor accelerating effect on the photo-degradation of PVC due to UV absorption, which was reflected in the degradation depth of 80 μm for PVC/wood composites as compared to 60 μm for neat PVC. Moreover, compounds based on polyene and carbonyl groups were observed on the exposed surfaces of the samples. Pillay *et al.* [111] reported that the flexural and impact properties of the nylon 6/carbon fibers composites remained stable towards UV radiations of 200-400 nm up to 600 h. However, the surface crystallinity of the composites increased to 44% from 40%, which was attributed to the molecular mobility and subsequent re-crystallization upon UV exposure.

Gu *et al.* [112] examined the changes in the surface and interfacial properties of poly(vinylidene fluoride) (PVDF)/poly(methyl methacrylate)-co-poly(ethyl acrylate) blend after treatment with UV radiations (275-800 nm) for 7 months at 50 °C and 9% RH. For the blends with lower content of PVDF, the photo-degradation rate was higher, which correspondingly resulted in rougher composite surfaces. On the other hand, the blends with higher content of PVDF presented only slight morphological changes after UV exposure. Effectively ordered microstructure of the composites was observed after UV treatment due to the acrylic copolymer erosion and corresponding rearrangement and re-crystallization of the PVDF molecular chains.

Islam *et al.* [113] demonstrated net decrease in the mechanical properties of the poly(lactic acid) (PLA) composites based on untreated hemp fibers as compared to the alkali treated hemp fibers based composites upon weathering using UV irradiation and water spray at 50 °C. During initial stages of weathering (up to 250 h), the untreated fibers based composites were more effective, whereas at prolonged weathering (500-1000 h), the alkali treated fibers based PLA composites showed higher weathering resistance. Flexural and tensile properties of both untreated and alkali treated fibers based composites decreased after weathering, whereas the impact strength of the composites was observed to increase. Leroux *et al.* [114] also evaluated the behavior of polystyrene (PS) nanocomposites based on 3-sulfopropyl methacrylate modified double layered hydroxide towards UV radiations of 300 nm. From the FTIR spectra, the filler was observed to have no effect on the UV accelerated degradation of PS. Also, a small increase in the oxidation rate was observed for the PS composite with 5 wt% filler.

Wochnowski *et al.* [115] examined the effect of UV wavelength (193, 248 and 308 nm) on the properties of poly(methyl methacrylate) (PMMA). For lower wavelength of 193 nm, crosslinking reaction between the ester side chains of the PMMA macromolecules was observed, which indicated further curing of PMMA. The application of UV radiations of 248 nm wavelength resulted in the splitting of the side chains from the backbone chain of PMMA, thus, creating more densification of the macromolecules with enhanced refractive index. The UV radiations of 308 nm wavelength resulted in the rupture of the polymer backbone and corresponding decomposition of the polymer. In another study, Caykara and Guven [116] also reported ester side chain rupture for PMMA and side chain rupture as well as crosslinking for copolymer of MMA and vinyltriethoxy silane in the presence of UV radiation of 259 nm wavelength.

Morlat-Therias *et al.* [117] investigated the influence of various MMTs (sodium form, intercalated and exfoliated) on the photo-degradation of ethylene-propylene-diene monomer (EPDM). Nano-composites with or without compatibilizer EPDM-g-MA were analyzed. The addition of MMTs had no influence on the photo-stabilization of EPDM, and the generation of similar photo-oxidation products was detected for all composites. EPDM/Na-MMT nano-composites exhibited almost identical induction period in the presence or absence of EPDM-g-MA, whereas a small reduction and disappearance was noticed for the corresponding intercalated and exfoliated MMT based nanocomposites respectively. The disappearance of the induction period was related to the presence organic cations in the exfoliated MMT, which accelerated the photo-degradation of EPDM. In another study, Morlat-Therias *et al.* [118] reported that the addition of photo-stabilizers such as Tinuvin P or 2-mercaptobenzimidazole hampered the photo-degradation of EPDM/MMT nanocomposites.

Photo-degradation of poly(styrene-*b*-(ethylene-*co*-butylene)-*b*-styrene) (SEBS) and the inhibiting effect of antioxidants were reported by Allen *et al.* [119]. At the time of UV exposure, photo-products such as anhydrides, peresters, peracids, α,β-unsaturated carbonyl products, etc., were generated due to the evolution of acetophenone and carboxylic acids on the corresponding styrene part and olefin end respectively. Also, hydroperoxidation was reported at the labile tertiary groups present on the butylene part. The effectiveness of antioxidants (phenolic and phosphite) to stop the action of acetophenone end groups was observed, which provided photostability to SEBS.

White *et al.* [120] reported the effect of natural and accelerated weathering on the photo-oxidative behavior of SEBS. For accelerated ageing, UV radiations of 295-450 nm were used under various conditions of temperature and relative humidity. Much faster photo-degradation of SEBS was reported when using a combination of different ageing parameters. The observed photo-degradation mechanisms under all weathering conditions were consistent for SEBS, while the photo-oxidation of natural weathered samples was not quick. A strong influence of temperature on the photo-degradation of the accelerated aged samples was noted, which was greater at 55 °C as compared to 30 °C. Also, the influence of relative humidity on the OH bond stretching was noticed.

1.3 Conclusions

The UV accelerated degradation and weathering of different polymers, blends and composites were reviewed in this chapter. Accelerated weathering and photo-degradation studies of polymers are needed to simulate their long term durability under various adverse external environmental conditions. The stability of polymeric materials towards weathering depends on many factors including the nature of polymers, added materials (micro- and nano-fillers/reinforcements, pigments, stabilizers, anti-degradants, etc.), presence of impurities and operating conditions such as wavelength, temperature, relative humidity, time of exposure, etc.

References

1. Pickett, J. (2001) Effect of environmental variables on the weathering of some engineering thermoplastics. *Polymer Preprints,* **42**, 424-425.
2. Pickett, J., Gibson, D., and Gardner, M. (2008) Effects of irradiation conditions on the weathering of engineering thermoplastics. *Polymer Degradation and Stability,* **93**, 1597-1606.
3. Heikkilä, A., Kärhä, P., Tanskanen, A., Kaunismaa, M., Koskela, T., Kaurola, J., Ture, T., and Syrjälä, S. (2009) Characterizing a UV chamber with mercury lamps for assessment of comparability to natural UV conditions. *Polymer Testing,* **28**, 57-65.
4. Larché, J.-F., Bussiere, P.-O., Therias, S., and Gardette, J.-L. (2012) Photooxidation of polymers: relating material properties to chemical changes. *Polymer Degradation and Stability,* **97**, 25-34.
5. Pandey, J. K., Reddy, K. R., Kumar, A. P., and Singh, R. P. (2005) An overview on the degradability of polymer nanocomposites. *Polymer Degradation and Stability,* **88**, 234-250.
6. Kumar, A. P., Depan, D., Tomer, N. S., and Singh, R. P. (2009) Nanoscale particles for polymer degradation and stabilization -

Trends and future perspectives. *Progress in Polymer Science,* **34,** 479-515.

7. Stark, N. M., and Matuana, L. M. (2004) Surface chemistry changes of weathered HDPE/wood-flour composites studied by XPS and FTIR spectroscopy. *Polymer Degradation and Stability,* **86,** 1-9.

8. Jabarin, S. A., and Lofgren, E. A. (1994) Photooxidative effects on properties and structure of high-density polyethylene. *Journal of Applied Polymer Science,* **53,** 411-423.

9. Fabiyi, J. S., McDonald, A. G., Wolcott, M. P., and Griffiths, P. R. (2008) Wood plastic composites weathering: Visual appearance and chemical changes. *Polymer Degradation and Stability,* **93,** 1405-1414.

10. Ruiz, C., and Machado, L. (2005) Accelerated weathering of UV/EB curable clearcoats. *Nuclear Instruments and Methods in Physics, Research Section B: Beam Interactions with Materials and Atoms,* **236,** 599-605.

11. Rabello, M., and White, J. (1997) The role of physical structure and morphology in the photodegradation behaviour of polypropylene. *Polymer Degradation and Stability,* **56,** 55-73.

12. Yakimets, I., Lai, D., and Guigon, M. (2004) Effect of photo-oxidation cracks on behaviour of thick polypropylene samples. *Polymer Degradation and Stability,* **86,** 59-67.

13. Nagai, N., Matsunobe, T., and Imai, T. (2005) Infrared analysis of depth profiles in UV-photochemical degradation of polymers. *Polymer Degradation and Stability,* **88,** 224-233.

14. Shyichuk, A., Stavychna, D., and White, J. (2001) Effect of tensile stress on chain scission and crosslinking during photo-oxidation of polypropylene. *Polymer Degradation and Stability,* **72,** 279-285.

15. Shyichuk, A., Turton, T., White, J., and Syrotynska, I. (2004) Different degradability of two similar polypropylenes as revealed by macromolecule scission and crosslinking rates. *Polymer Degradation and Stability,* **86,** 377-383.

16. Guadagno, L., Naddeo, C., and Vittoria, V. (2004) Structural and morphological changes during UV irradiation of the crystalline helical form of syndiotactic polypropylene. *Macromolecules,* **37,** 9826-9834.

17. Tocháček, J., and Vrátníčková, Z. (2014) Polymer life-time prediction: The role of temperature in UV accelerated ageing of polypropylene and its copolymers. *Polymer Testing,* **36,** 82-87.

18. Francois-Heude, A., Richaud, E., Desnoux, E., and Colin, X. (2014) Influence of temperature, UV-light wavelength and intensity on polypropylene photothermal oxidation. *Polymer Degradation and Stability,* **100,** 10-20.

19. Butylina, S., Hyvärinen, M., and Kärki, T. (2012) Accelerated weathering of wood–polypropylene composites containing miner-

als. *Composites, Part A: Applied Science and Manufacturing*, **43**, 2087-2094.

20. Kallakas, H., Poltimäe, T., Süld, T.-M., Kers, J., and Krumme, A. (2015) The influence of accelerated weathering on the mechanical and physical properties of wood-plastic composites. *Proceedings of the Estonian Academy of Sciences*, **64**, doi: 10.3176/proc.2015.1S.05.

21. La Mantia, F. P., and Morreale, M. (2008) Accelerated weathering of polypropylene/wood flour composites. *Polymer Degradation and Stability*, **93**, 1252-1258.

22. Beg, M. D. H., and Pickering, K. L. (2008) Accelerated weathering of unbleached and bleached Kraft wood fibre reinforced polypropylene composites. *Polymer Degradation and Stability*, **93**, 1939-1946.

23. Butylina, S., Hyvärinen, M., and Kärki, T. (2012) A study of surface changes of wood-polypropylene composites as the result of exterior weathering. *Polymer Degradation and Stability*, **97**, 337-345.

24. Peng, Y., Liu, R., Cao, J., and Chen, Y. (2014) Effects of UV weathering on surface properties of polypropylene composites reinforced with wood flour, lignin, and cellulose. *Applied Surface Science*, **317**, 385-392.

25. Peng, Y., Liu, R., and Cao, J. (2015) Characterization of surface chemistry and crystallization behavior of polypropylene composites reinforced with wood flour, cellulose, and lignin during accelerated weathering. *Applied Surface Science*, **332**, 253-259.

26. Fabiyi, J. S., and McDonald, A. G. (2014) Degradation of polypropylene in naturally and artificially weathered plastic matrix composites. *Maderas: Ciencia y Tecnología*, **16**, 275-290.

27. Darie, R. N., Bodirlau, R., Teaca, C. A., Macyszyn, J., Kozlowski, M., and Spiridon, I. (2013) Influence of accelerated weathering on the properties of polypropylene/polylactic acid/eucalyptus wood composites. *International Journal of Polymer Analysis and Characterization*, **18**, 315-327.

28. Abu-Sharkh, B., and Hamid, H. (2004) Degradation study of date palm fibre/polypropylene composites in natural and artificial weathering: mechanical and thermal analysis. *Polymer Degradation and Stability*, **85**, 967-973.

29. Thwe, M. M., and Liao, K. (2002) Effects of environmental aging on the mechanical properties of bamboo–glass fiber reinforced polymer matrix hybrid composites. *Composites, Part A: Applied Science and Manufacturing*, **33**, 43-52.

30. Joseph, P., Rabello, M. S., Mattoso, L., Joseph, K., and Thomas, S. (2002) Environmental effects on the degradation behaviour of sisal fibre reinforced polypropylene composites. *Composites Science and Technology*, **62**, 1357-1372.

31. Gadioli, R., Morais, J. A., Waldman, W. R., and De Paoli, M.-A. (2014)

The role of lignin in polypropylene composites with semi-bleached cellulose fibers: Mechanical properties and its activity as antioxidant. *Polymer Degradation and Stability,* **108**, 23-34.

32. Thattaiparthasarathy, K. B., Pillay, S., Ning, H., and Vaidya, U. (2008) Process simulation, design and manufacturing of a long fiber thermoplastic composite for mass transit application. *Composites, Part A: Applied Science and Manufacturing,* **39**, 1512-1521.

33. Markarian, J. (2007) Long fibre reinforced thermoplastics continue growth in automotive. *Plastics, Additives and Compounding,* **9**, 20-24.

34. Goel, A., Chawla, K., Vaidya, U., Koopman, M., and Dean, D. (2008) Effect of UV exposure on the microstructure and mechanical properties of long fiber thermoplastic (LFT) composites. *Journal of Materials Science,* **43**, 4423-4432.

35. Chevali, V. S., Dean, D. R., and Janowski, G. M. (2010) Effect of environmental weathering on flexural creep behavior of long fiber-reinforced thermoplastic composites. *Polymer Degradation and Stability,* **95**, 2628-2640.

36. Rabello, M., Tocchetto, R., Barros, L., d'Almeida, J., and White, J. (2013) Weathering of polypropylene composites containing weldlines. *Plastics, Rubber and Composites,* **30(**3), 132-140.

37. Zhu, J., Morgan, A. B., Lamelas, F. J., and Wilkie, C. A. (2001) Fire properties of polystyrene-clay nanocomposites. *Chemistry of Materials,* **13**, 3774-3780.

38. Bocchini, S., Morlat-Thérias, S., Gardette, J.-L., and Camino, G. (2007) Influence of nanodispersed boehmite on polypropylene photooxidation. *Polymer Degradation and Stability,* **92**, 1847-1856.

39. Tidjani, A., and Wilkie, C. A. (2001) Photo-oxidation of polymeric-inorganic nanocomposites: chemical, thermal stability and fire retardancy investigations. *Polymer Degradation and Stability,* **74**, 33-37.

40. Mailhot, B., Morlat, S., Gardette, J.-L., Boucard, S., Duchet, J., and Gerard, J.-F. (2003) Photodegradation of polypropylene nanocomposites. *Polymer Degradation and Stability,* **82**, 163-167.

41. Bocchini, S. Morlat-Therias, S. Gardette, J. L. and Camino, G. (2008) Influence of nanodispersed hydrotalcite on polypropylene photooxidation. *European Polymer Journal,* **44**, 3473-3481.

42. Alongi, J., Poskovic, M., Frache, A., and Trotta, F. (2011) Role of β-cyclodextrin nanosponges in polypropylene photooxidation. *Carbohydrate Polymers,* **86**, 127-135.

43. Kumanayaka, T., Jollands, M., and Parthasarathy, R. (2009) The Effect of Nanoclay on Photo-oxidation of Polyethylene. *8th World Congress of Chemical Engineering (WCCE8),* Canada.

44. Dehbi, A., Bouaza, A., Hamou, A., Youssef, B., and Saiter, J. M. (2010) Artificial ageing of tri-layer polyethylene film used as greenhouse

cover under the effect of the temperature and the UV-A simultaneously. *Materials and Design,* **31**, 864-869.

45. Dehbi, A., Mourad, A.-H. I., and Bouaza, A. (2011) Ageing effect on
the properties of tri-layer polyethylene film used as greenhouse
roof. *Procedia Engineering,* **10**, 466-471.

46. Dehbi, A., Mourad, A. H. I., Djakhdane, K., and Hilal-Alnaqbi, A.
(2015) Degradation of thermomechanical performance and life-
time estimation of multilayer greenhouse polyethylene films under
simulated climatic conditions. *Polymer Engineering and Science,*
55, 287-298.

47. Imane, B. M., Asma, A., Fouad, C. S., and Mohamed, S. (2015)
Weathering effects on the microstructure morphology of low density polyethylene. *Procedia - Social and Behavioral Sciences,* **195**,
2228-2235.

48. Mohamed, F. H., Mourad, A.-H., and Barton, D. (2013) UV irradiation and aging effects on nanoscale mechanical properties of ultra
high molecular weight polyethylene for biomedical implants. *Plastics, Rubber and Composites,* **31**(8), 346-352.

49. Stark N. M., and Matuana, L. M. (2007) Characterization of weathered wood–plastic composite surfaces using FTIR spectroscopy,
contact angle, and XPS. *Polymer Degradation and Stability,* **92**,
1883-1890.

50. Lundin, T., Cramer, S. M., Falk, R. H., and Felton, C. (2004) Accelerated weathering of natural fiber-filled polyethylene composites.
Journal of Materials in Civil Engineering, **16**, 547-555.

51. Stark, N. M., and Matuana, L. M. (2006) Influence of photostabi-
lizers on wood flour-HDPE composites exposed to xenon-arc radiation with and without water spray. *Polymer Degradation and Stability,* **91**, 3048-3056.

52. Fabiyi, J. S., McDonald, A. G., and McIlroy, D. (2009) Wood modification effects on weathering of HDPE-based wood plastic composites. *Journal of Polymers and the Environment,* **17**, 34-48.

53. Fabiyi, J. S., and McDonald, A. G. (2010) Effect of wood species on
property and weathering performance of wood plastic composites.
Composites, Part A: Applied Science and Manufacturing, **41**, 1434-
1440.

54. Krehula, L. K., Katančić, Z., Siročić, A. P., and Hrnjak-Murgić, Z.
(2014) Weathering of high-density polyethylene-wood plastic
composites. *Journal of Wood Chemistry and Technology,* **34**, 39-54.

55. Kumanayaka, T., Parthasarathy, R., and Jollands, M. (2010) Accelerating effect of montmorillonite on oxidative degradation of polyethylene nanocomposites. *Polymer Degradation and Stability,* **95**,
672-676.

56. Botta, L., Dintcheva, N. T., and La Mantia, F. P. (2009) The role of
organoclay and matrix type in photo-oxidation of polyolefin/clay

nanocomposite films. *Polymer Degradation and Stability,* **94**, 712-718.

57. Dintcheva, N. T., Al-Malaika, S., and La Mantia, F. P. (2009) Effect of extrusion and photo-oxidation on polyethylene/clay nanocomposites. *Polymer Degradation and Stability,* **94**, 1571-1588.

58. Grigoriadou, I., Paraskevopoulos, K., Chrissafis, K., Pavlidou, E., Stamkopoulos, T.-G., and Bikiaris, D. (2011) Effect of different nanoparticles on HDPE UV stability. *Polymer Degradation and Stability,* **96**, 151-163.

59. Dintcheva, N. T., La Mantia, F., and Malatesta, V. (2009) Photo-oxidation behaviour of polyethylene/multi-wall carbon nanotube composite films. *Polymer Degradation and Stability,* **94**, 162-170.

60. Grigoriadou, I., Paraskevopoulos, K., Karavasili, M., Karagiannis, G., Vasileiou, A., and Bikiaris, D. (2013) HDPE/Cu-nanofiber nanocomposites with enhanced mechanical and UV stability properties. *Composites, Part B: Engineering,* **55**, 407-420.

61. Yang, R., Christensen, P., Egerton, T., and White, J. (2010) Degradation products formed during UV exposure of polyethylene-ZnO nano-composites. *Polymer Degradation and Stability,* **95**, 1533-1541.

62. Saleem, H., Edathil, A., Ncube, T., Pokhrel, J., Khoori, S., Abraham, A., and Mittal, V. (2016) Mechanical and thermal properties of thermoset-graphene nanocomposites. *Macromolecular Materials and Engineering,* **301**, 231-259.

63. Malshe, V., and Waghoo, G. (2004) Chalk resistant epoxy resins. *Progress in Organic Coatings,* **51**, 172-180.

64. Malshe, V., and Waghoo, G. (2004) Weathering study of epoxy paints. *Progress in Organic Coatings,* **51**, 267-272.

65. Islam, M., Pickering, K., and Foreman, N. (2011) The effect of accelerated weathering on the mechanical properties of alkali treated hemp fibre/epoxy composites. *Journal of Adhesion Science and Technology,* **25**, 1947-1959.

66. Dixit, S., and Verma, P. (2014) Effect of UV Exposure on Coir-epoxy Infused Particulate Composite. *American Journal of Polymer Science & Engineering,* **2**, 1-11.

67. Yan, L., Chouw, N., and Jayaraman, K. (2015) Effect of UV and water spraying on the mechanical properties of flax fabric reinforced polymer composites used for civil engineering applications. *Materials and Design,* **71**, 17-25.

68. Dubois, C., Monney, L., Bonnet, N., and Chambaudet A. (1999) Degradation of an epoxy-glass-fibre laminate under photo-oxidation/leaching complementary constraints. *Composites, Part A: Applied Science and Manufacturing,* **30**, 361-368.

69. Kumar, B. G., Singh, R. P., and Nakamura, T. (2002) Degradation of carbon fiber-reinforced epoxy composites by ultraviolet radiation

and condensation. *Journal of Composite Materials,* **36**, 2713-2733.

70. Singh, A. K., and Singh, R. P. (2006) Effect of Mechanical Loading and Environmental Degradation on Carbon Fiber Reinforced Composites. *Proceedings of the 2006 SEM Annual Conference and Exposition on Experimental and Applied Mechanics,* USA.

71. Awaja, F., and Pigram, P. J. (2009) Surface molecular characterisation of different epoxy resin composites subjected to UV accelerated degradation using XPS and ToF-SIMS. *Polymer Degradation and Stability,* **94**, 651-658.

72. Awaja, F., Nguyen, M.-T., Zhang, S., and Arhatari, B. (2011) The investigation of inner structural damage of UV and heat degraded polymer composites using X-ray micro CT. *Composites, Part A: Applied Science and Manufacturing,* **42**, 408-418.

73. Chang, L., and Chow, W. (2010) Accelerated weathering on glass fiber/epoxy/organo-montmorillonite nanocomposites. *Journal of Composite Materials,* **44**(12), 1421-1434.

74. Zainuddin, S., Hosur, M., Barua, R., Kumar, A., and Jeelani, S. (2011) Effects of ultraviolet radiation and condensation on static and dynamic compression behavior of neat and nanoclay infused epoxy/glass composites. *Journal of Composite Materials,* **45**, 1901-1918.

75. Woo, R. S., Chen, Y., Zhu, H., Li, J., Kim, J.-K., and Leung, C. K. (2007) Environmental degradation of epoxy–organoclay nanocomposites due to UV exposure. Part I: Photo-degradation. *Composites Science and Technology,* **67**, 3448-3456.

76. Woo, R. S., Zhu, H., Leung, C. K., and Kim, J.-K. (2008) Environmental degradation of epoxy-organoclay nanocomposites due to UV exposure: Part II residual mechanical properties. *Composites Science and Technology,* **68**, 2149-2155.

77. Mailhot, B., Morlat-Therias, S., Bussiere, P.-O., Le Pluart, L., Duchet, J., Sautereau, H., Gérard, J.-F., and Gardette, J.-L. (2008) Photoageing behaviour of epoxy nanocomposites: Comparison between spherical and lamellar nanofillers. *Polymer Degradation and Stability,* **93**, 1786-1792.

78. Shanmugam, N., Hussein, I. A., Badghaish, A., Shuaib, A. N., Furquan, S. A., and Al-Mehthel, M. H. (2015) Evaluation of oil fly ash as a light stabilizer for epoxy composites: Accelerated weathering study. *Polymer Degradation and Stability,* **112**, 94-103.

79. Haq, M. I. U. (2007) Applications of unsaturated polyester resins. *Russian Journal of Applied Chemistry,* **80**, 1256-1269.

80. Ebewele, R. O. (2000) *Polymer Science and Technology,* CRC Press, USA.

81. Jia, Z., Li, X., and Zhao, Q. (2010) Effect of artificial weathering on surface properties of unsaturated polyester (UP) resin. *Materials Chemistry and Physics,* **121**, 193-197.

82. Gu, H. (2008) Degradation of glass fibre/polyester composites after ultraviolet radiation. *Materials and Design,* **29**, 1476-1479.

83. Hongwang, Q., and Huang, G. (2011) Mechanical behaviors of glass/polyester composites after UV radiation. *Journal of Composite Materials,* **45**, 1939-1943.

84. Mouzakis, D. E., Zoga, H., and Galiotis, C. (2008) Accelerated environmental ageing study of polyester/glass fiber reinforced composites (GFRPCs). *Composites, Part B: Engineering,* **39**, 467-475.

85. Rodrigues, L., Silva, R., and Aquino, E. (2012) Effect of accelerated environmental aging on mechanical behavior of curaua/glass hybrid composite. *Journal of Composite Materials,* **46**, 2055-2064.

86. Shahzad, A. (2014) Accelerated weathering properties of hemp fibre composites. *Open Access Library Journal,* **1**, 1-8.

87. Chattopadhyay, D., and Raju, K. (2007) Structural engineering of polyurethane coatings for high performance applications. *Progress in Polymer Science,* **32**, 352-418.

88. Yang, X., Vang, C., Tallman, D., Bierwagen, G., Croll, S., and Rohlik, S. (2001) Weathering degradation of a polyurethane coating. *Polymer Degradation and Stability,* **74**, 341-351.

89. Yang, X. F., Li, J., Croll, S., Tallman, D., and Bierwagen, G. (2003) Degradation of low gloss polyurethane aircraft coatings under UV and prohesion alternating exposures. *Polymer Degradation and Stability,* **80**, 51-58.

90. Rosu, D., Rosu, L., and Cascaval, C. N. (2009) IR-change and yellowing of polyurethane as a result of UV irradiation. *Polymer Degradation and Stability,* **94**, 591-596.

91. Boubakri, A., Guermazi, N., Elleuch, K., and Ayedi, H. (2010) Study of UV-aging of thermoplastic polyurethane material. *Materials Science and Engineering A,* **527**, 1649-1654.

92. Saadat-Monfared, A., and Mohseni, M. (2014) Polyurethane nanocomposite films containing nano-cerium oxide as UV absorber; Part 2: Structural and mechanical studies upon UV exposure. *Colloids and Surfaces A: Physicochemical and Engineering Aspects,* **441**, 752-757.

93. Zia, K. M., Barikani, M., Bhatti, I. A., Zuber, M., and Barmar, M. (2009) XRD studies of UV-irradiated chitin based polyurethane elastomers. *Carbohydrate Polymers,* **77**, 54-58.

94. Zia, K. M., Barikani, M., Khalid, A. M., and Honarkar, H. (2009) Surface characteristics of UV-irradiated chitin-based polyurethane elastomers. *Carbohydrate Polymers,* **77**, 621-627.

95. Zia, K. M., Zuber, M., Mahboob, S., Sultana, T., and Sultana, S. (2010) Surface characteristics of UV-irradiated chitin-based shape memory polyurethanes. *Carbohydrate Polymers,* **80**, 229-234.

96. Fried, J. R. (2014) *Polymer Science and Technology*, Pearson Education, USA.

97. Tjandraatmadja, G., Burn, L., and Jollands, M. (2002) Evaluation of commercial polycarbonate optical properties after QUV-A radiation - the role of humidity in photodegradation. *Polymer Degradation and Stability*, **78**, 435-448.

98. Migahed, M., and Zidan, H. (2006) Influence of UV-irradiation on the structure and optical properties of polycarbonate films. *Current Applied Physics*, **6**, 91-96.

99. Pérez, J. M., Vilas, J. L., Laza, J. M., Arnáiz, S., Mijangos, F., Bilbao, E., Rodríguez, M., and León, L. M. (2010) Effect of reprocessing and accelerated ageing on thermal and mechanical polycarbonate properties. *Journal of Materials Processing Technology*, **210**, 727-733.

100. Sloan J. M., and Patterson, P. (2005) *Mechanisms of Photo Degradation for Layered Silicate-Polycarbonate Nanocomposites*, Army Research Laboratory, USA. Online: https://apps.dtic.mil/dtic/tr/fulltext/u2/a441212.pdf [accessed 19th March 2019].

101. Da Silva, A. N., Teixeira, S., Widal, A., and Coutinho, F. (2001) Mechanical properties of polymer composites based on commercial epoxy vinyl ester resin and glass fiber. *Polymer Testing*, **20**, 895-899.

102. Rosu, D., Rosu, L., and Cascaval, C. N. (2008) Effect of ultraviolet radiation on vinyl ester network based on bisphenol A. *Journal of Photochemistry and Photobiology A: Chemistry*, **194**, 275-282.

103. Signor, A. W., VanLandingham, M. R., and Chin, J. W. (2003) Effects of ultraviolet radiation exposure on vinyl ester resins: characterization of chemical, physical and mechanical damage. *Polymer Degradation and Stability*, **79**, 359-368.

104. Rosu, L., Cascaval, C. N., and Rosu, D. (2009) Effect of UV radiation on some polymeric networks based on vinyl ester resin and modified lignin. *Polymer Testing*, **28**, 296-300.

105. Korach C., and Chiang, F. (2012) Characterization of Carbon Fiber-Vinylester Composites Exposed to Combined UV Radiation and Salt Spray. *15th European Conference on Composite Materials*, Italy.

106. Hassan, C. M., and Peppas, N. A. (2000) Structure and applications of poly (vinyl alcohol) hydrogels produced by conventional cross-linking or by freezing/thawing methods. *Advances in Polymer Science*, **153**, 37-65.

107. Kaczmarek, H., and Podgórski, A. (2007) The effect of UV-irradiation on poly (vinyl alcohol) composites with montmorillonite. *Journal of Photochemistry and Photobiology A: Chemistry*, **191**, 209-215.

108. Kaczmarek, H., and Podgórski, A. (2007) Photochemical and thermal behaviours of poly (vinyl alcohol)/graphite oxide composites. *Polymer Degradation and Stability*, **92**, 939-946.

109. Moon, Y.-E., Yun, J., Kim, H.-I., and Lee, Y.-S. (2011) Effect of graphite oxide on photodegradation behavior of poly (vinyl alcohol)/graphite oxide composite hydrogels. *Carbon Letters*, **12**, 138-142.

110. Chaochanchaikul, K., Rosarpitak, V., and Sombatsompop, N. (2013) Photodegradation profiles of PVC compound and wood/PVC composites under UV weathering. *Express Polymer Letters*, **7**, 146-160.

111. Pillay, S., Vaidya, U. K., and Janowski, G. M. (2009) Effects of moisture and UV exposure on liquid molded carbon fabric reinforced nylon 6 composite laminates. *Composites Science and Technology*, **69**, 839-846.

112. Gu, X., Michaels, C., Nguyen, D., Jean, Y., Martin, J., and Nguyen, T. (2006) Surface and interfacial properties of PVDF/acrylic copolymer blends before and after UV exposure. *Applied Surface Science*, **252**, 5168-5181.

113. Islam, M. S., Pickering, K. L., and Foreman, N. J. (2010) Influence of accelerated ageing on the physico-mechanical properties of alkali-treated industrial hemp fibre reinforced poly (lactic acid)(PLA) composites. *Polymer Degradation and Stability*, **95**, 59-65.

114. Leroux, F., Meddar, L., Mailhot, B., Morlat-Thérias, S., and Gardette, J.-L. (2005) Characterization and photooxidative behaviour of nanocomposites formed with polystyrene and LDHs organo-modified by monomer surfactant. *Polymer*, **46**, 3571-3578.

115. Wochnowski, C., Eldin, M. S., and Metev, S. (2005) UV-laser-assisted degradation of poly (methyl methacrylate). *Polymer Degradation and Stability*, **89**, 252-264.

116. Caykara T., and Güven O. (1999) UV degradation of poly (methyl methacrylate) and its vinyltriethoxysilane containing copolymers. *Polymer Degradation and Stability*, **65**, 225-229.

117. Morlat-Therias, S., Mailhot, B., Gardette, J.-L., Da Silva, C., Haidar, B., and Vidal, A. (2005) Photooxidation of ethylene-propylene-diene/montmorillonite nanocomposites. *Polymer Degradation and Stability*, **90**, 78-85.

118. Morlat-Therias, S., Fanton, E., Tomer, N. S., Rana, S., Singh, R., and Gardette, J.-L., (2006) Photooxidation of vulcanized EPDM/montmorillonite nanocomposites. *Polymer Degradation and Stability*, **91**, 3033-3039.

119. Allen, N. S., Luengo, C., Edge, M., Wilkinson, A., Parellada, M. D., Barrio, J. A., and Santa Quiteria, V. R. (2004) Photooxidation of styrene-ethylene-butadiene-styrene (SEBS) block copolymer. *Journal of Photochemistry and Photobiology A: Chemistry*, **162**, 41-51.

120. White, C., Tan, K., Hunston, D., Nguyen, T., Benatti, D., Stanley, D., and Chin, J. (2011) Laboratory accelerated and natural weathering of styrene-ethylene-butylene-styrene (SEBS) block copolymer. *Polymer Degradation and Stability*, **96**, 1104-1110.

2

UV Degradation of Polymer Coatings

2.1 Introduction

Polymeric coatings are developed for meeting various needs, such as corrosion protection, moisture and gas barrier, maintenance of equipment and surfaces, etc. The protective polymeric coatings are of substantial importance in present-day technologies, and extensive research efforts have been carried out for the advancement of such coatings with enhanced environmental stability. Unlike the architectural coatings, which are largely used in amiable weather conditions, the protective coatings are usually applied in challenging environments. As an illustration, the pipelines with surface coatings, employed for the transportation of oil, gas or chemicals over longer distances, are subjected to severe hot, cold, arid and humid conditions.

Organic coatings are extensively employed for preventing the corrosion of the metallic structures due to their effective properties, low manufacturing cost, versatility, ease of application and aesthetic features [1]. Different physical processes as well as chemical reactions take place when the synthetic or natural materials based coatings are exposed to outdoor conditions for prolonged periods of time. Such physical processes and chemical reactions are collectively termed as weathering [2]. The exterior durability of an organic coating is described as the resistance offered by the material to the unwanted effects induced by the natural environment to which the coating is exposed to during its service life [3]. The longer a coating is capable of avoiding failure because of its good weathering resistance, the better is its durability. Service lifetime is defined as the time that a coating can last till it fails. The service lifetime of a polymer coating is determined by the coating characteristics as well as the service conditions in which the coating is located. Overall, the development of coatings with superior weathering resistance, along with other property specifications, is an important research area.

Haleema Saleem and Vikas Mittal*, The Petroleum Institute (part of Khalifa University of Science and Technology), Abu Dhabi, UAE
*Current address: Bletchington, Wellington County, Australia

Various studies have been carried out to understand the degradation behavior of the polymer coatings in the past several years. The ultraviolet (UV) radiation, O_2 and H_2O are the three discriminating factors responsible for the degradation of coatings during weathering [4]. When the aforementioned factors are united with several other environmental variables like wind and seasonal periodicity, the achievement of weather resistant coatings becomes even more challenging.

The UV light is an electromagnetic radiation having wavelength between 10-400 nm and energy ranging from 3-124 eV [5]. As per the ISO solar irradiance standard (ISO 21348), the electromagnetic spectrum of UV is subdivided into the following three major groups [5,6]:

- Ultraviolet A (UVA): These radiations have wavelength in the range 320-400 nm. Almost 99% of the total UV light reaching the earth's surface is UVA. It is accountable for several photosensitivity reactions, and it can enhance the adverse effects of ultraviolet B radiations.
- Ultraviolet B (UVB): Nearly 1% of the total UV light reaching the earth's surface is UVB. These radiations have wavelength in the range 290-320 nm. UVB generates several detrimental photo-chemical reactions.
- Ultraviolet C (UVC): These radiations have wavelength in the range 200-290 nm. Usually, UVC cannot reach the earth's surface due to the blockage by the ozone layer.

For majority of the synthetic polymers, the most substantial degradation mechanisms are linked to the absorption of UV light having energy between 300-450 kJ/mol [7]. When the quantity of energy absorbed by the polymer outstrips the polymer bond energy, the UV degradation takes place [7]. The rate of polymer degradation depends on various factors like the exposure location, temperature, nature of the substrate, polymer material, etc. The degradation of the polymeric coatings can be established in the form of swelling, color variation, crosslinking, oxidation, water absorption and dissolution [6]. Further, at elevated temperatures, certain gas species may be generated from the coatings, thereby, varying the glass transition temperature, molecular weight, gloss and density of the polymer matrix in the coatings. These factors generally lead to an increment in the brittleness and porosity of the polymeric coatings [8,9]. The photo-degradation of the crosslinked (thermosetting) polymer coatings generally also causes variations in the crosslink density that can change the glass transition temperature. Thus, the recognition of photo-degradation is connected to realizing the physical and chemical effects as well as the relations between them. The photo-degradation processes in the top layers of the polymeric coatings are spatially non-uniform in nature because of the nature of UV absorptivity of the polymeric materials and ingress of water and oxygen. Prominent examples of the physical properties of the polymer coatings affected by the photo-degradation processes include oxygen permeability [10], optical properties (color, UV-vis absorption) [11,12], water sorption [13], gloss [14,15] , fracture energy [16] , surface tension [14,15], hardness [17], internal stresses [18], elastic modulus [19], crosslink density [18] and glass transition temperature [20]. Conventional failure modes are related to the coating appearance as well as its mechanical integrity (delamination and cracking failure). The failure associated with the crack formation is due to the fact that the coatings turn brittle at the time of degradation and both quick fatigue failure and brittle failure may take place [21]. Various degradation factors affect the failure modes in distinct ways and each failure mode might exhibit a diversified dependence on particular coating feature, e.g. the layer thickness [16].

Due to weathering, the service life of polymers in the open air applications becomes limited. The service life of the polymeric coatings can be considerably lengthened, either by the surface treatment to screen the adverse UV radiations or by the incorporation of light stabilizers such as hindered amine light stabilizer (HALS), UV absorbers, etc. [22]. For the development of polymeric coatings with improved properties, characterization of the long term degradation is also a fundamental necessity, and various techniques have been employed in this respect. For instance, several techniques address the macroscopic properties of the system such as weight, mechanical integrity, loss of gloss and variation in contact angle [23]. These techniques provide information associated with the material performance [24], however, these do not give insights about the atomic scale degradation. On the other hand, Fourier transform infrared (FTIR) spectroscopy, scanning electron microscopy (SEM), electron spin resonance (ESR), Raman spectroscopy, nuclear magnetic resonance (NMR) and positron annihilation spectroscopy (PAS) are extensively used for analyzing the microscopic as well as atomic scale degradation [25,26]. Out of these, PAS and ESR are specifically employed for identifying the early stages of coating degradation.

An improved understanding of the degradation behavior supports the forecasting of the performance of the polymer coatings, along with tuning of the coatings formulations for enhancing their durability. In this respect, the photo-degradation of diverse polymer coatings (e.g. acrylic, epoxy, polyurethane (PU), silicone and polyester coatings) has been discussed in detail in the current chapter.

2.2 Mechanism of UV Degradation

The UV degradation of the polymer materials is generally induced by a complex series of chemical reactions, which are brought about by the UV light absorption. The physical properties of the polymer deteriorate as a result of the UV degradation, which is one of the most common issues seen in the polymeric coatings exposed to sunlight. The continuous exposure to UV light causes more severe damage than the intermittent exposure. Based on the changes occurring in the polymers under the effect of UV radiation exposure, the synthetic polymers might be divided into two groups. The first group consists of the polymers that discolor very quickly when exposed to UV radiation, however, maintain the physical properties for continued periods of irradiation. Due to the UV radiation exposure, changes occur in the chemical structure, thereby, activating the chromophoric groups. Nevertheless, the scission of polymer backbone does not occur in these type of polymers. Polyacrylonitrile (PAN) and polyvinylchloride (PVC) are examples of these types of polymers [27]. The second type of polymers are those which become brittle during the exposure of UV radiation. This is due to the breaking of the main chains as well as photo-induced crystallization [6]. Polymers such as polyethylene, polystyrene and polypropylene are the examples of this category. The UV light of high intensity causes the generation of free radicals on the surface of the polymer [28]. Hence, this initiates the crosslinking reactions for the additional polymerization, oxidation or both [5]. These radicals can easily react with the oxygen present in the air, thereby, enabling auto-accelerating photo-oxidation. Environmental conditions like temperature, humidity, acidity/basicity, pollutants and oxygen remarkably enhance the UV degradation level [29]. The presence of water is very crucial because of its direct participation during the degradation process. The water molecules may also lead to the matrix plasticization that can change the polymer coating's effective glass transition temperature, solubility and diffusion coefficient of additional degradation agents like oxygen [30].

Similar to the radical polymerization chemistry, the photo-degradation process is defined in terms of initiation, propagation and termination reactions [31]. The initiation process includes a photolytic scission that takes place after the UV absorption by the polymer molecules, thus, causing the chain scission into two radicals. The polymer radical is effortlessly oxidized to obtain a peroxy radical, which can abstract a hydrogen atom deriving from a different polymer fragment in the coating to obtain hydroperoxide as well as additional polymer radical. The hydroperoxide decomposes to form a polymer oxy-radical as well as a hydroxy radical under the influence of temperature and photons. Both radical species take part in the hydrogen abstraction processes and form new polymer radicals, which can again take part in the oxidation reactions. Further, the polymer oxy-radical present in the polymer chains can cause chain scission. Finally, the terminal reactions occur via radical recombination process.

As mentioned earlier, the durability of the polymer coatings can be enhanced by the incorporation of stabilizers in the formulations. These stabilizers have the ability to target the initiation as well as propagation reactions during the photo-degradation process. The initiation reactions can be limited by avoiding a part of the incident UV radiation from being absorbed by the polymer. This can be accomplished by the reflection and absorption of the incident radiation as well as the consequent dissipation of its energy as heat and long wavelength radiation [32]. Apart from avoiding the absorption process, the initiation process can also be avoided after the development of an absorption event, in case the excited chromophore energy is moved to a stabilizer molecule before the chain breaking occurs. This method is known as the quenching of the excited states and can be obtained by the incorporation of metal chelate compounds to the polymeric coatings [33]. Not every UV photon can be counterbalanced by the stabilization mechanisms, thus, the initiation might take place partially. Due to this reason, radical scavengers might be incorporated in the coatings formulations for partially avoiding the propagation processes. The most widely employed radical scavenger is HALS, which is based on nitroxy radicals ($R-NO^{•}$). The nitroxy radicals recombine with the polymer radicals to generate alkyloxyamine. The selection of additives that constitute the stabilizing system has to be tuned on the basis of the photo-degradation mechanism of the particular polymer needed to be stabilized. Hence, an excellent knowledge of the photo-degradation mechanisms of the polymeric materials is essential for effective stabilization.

2.3 Different Polymer Coatings Systems

Polymer resins are the starting component of all coating formulations and inevitably guide the fundamental properties of the coatings systems. Unlike the architectural coatings where the water borne coatings systems dominate, the protective coatings dominantly rely on the solvent borne systems, as these facilitate the usage of highly durable and high T_g resins. The UV degradation performance of different polymeric coatings based on acrylics, epoxies, PUs, silicones and polyesters is discussed in the following sections.

2.3.1 PU Coatings

PUs are employed in an extensive variety of technical as well as commercial applications, due to their chemical resistance, good mechanical properties, processability and high tensile strength [34]. PU coatings are commonly utilized for the protection of materials such as metals, wood and plastics because of their color retention, excellent gloss and dimensional stability [35]. One of the important drawbacks of PU based coatings is severe light sensitivity, specifically UV light. This limits the application of PU materials as the surface coatings for external utilization. When exposed to high energy UV radiations, PU goes through remarkable structural variation, which leads to the deterioration in its physical properties [36,37]. PU materials prepared using aromatic isocyanates exhibit yellowing, when polymer coating is exposed to UV light, due to the oxidation process occurring in the polymer backbone. The stabilization against high energy radiations can be achieved by the addition of photo-stabilizers and anti-oxidants [36]. Zinc oxide (ZnO) and titanium dioxide (TiO_2) have been observed to be effective additives that promote good photo-stabilization effect in PUs.

Yang *et al.* [4] analyzed the degradation of epoxy primer with high gloss PU topcoat coating system, which was exposed either in a prohesion chamber and a QUV chamber alternatively, or only in a QUV chamber. The atomic force microscopy (AFM) analysis confirmed the formation of micro-blisters on the coating surface after both exposures. During the QUV exposure, the blisters increased in size on enhancing the exposure time. However, in the case of prohesion/QUV alternative exposure, the blisters maintained their size throughout the time of exposure. The surface roughness (RMS) exhibited a steady enhancement with the time of QUV exposure, but was maintained constant during the prohesion/QUV exposure. The SEM analysis proved that the exposure to QUV was more destructive than the pro-hesion/QUV alternative exposure. In a dry/wet interchanging envi-ronment, the blisters with sub-micrometer to micrometer dimen-sions were formed on the coating surface.

He *et al.* [23] examined the photo-degradation of PU coatings by the electron spin resonance (ESR) method. The PU film specimens were subjected to various accelerated ageing conditions, which in-cluded narrow band UV irradiation of 340 and 313 nm, increased temperature without any irradiation and broadband irradiation from a xenon arc lamp. In addition, the effect of titania on the free radical generation was also investigated. It was observed that the radicals decayed in a period of 1 hour following the irradiation, in the pres-ence of ambient oxygen. However, in the absence of oxygen, two long lifetimes of the order of 60 and 350 hours were detected. In another study, Singh *et al.* [38] reported a comprehensive yellowing of the urethane clear coatings based on castor oil and diphenylmethane diisocyanate (MDI) in the presence of sunlight. The clear aromatic PU coatings were attained by combining MDI and castor oil in 1:1 ratio at room temperature and subsequently applying on glass plate. To provide light stability and to restrain the yellowing in aromatic PU coatings, modification with stabilizers as well as their synergistic mixtures was suggested by Decker and Bandaikha [39].

2.3.2 Acrylic Coatings

The acrylic/methacrylic polymers have been widely employed in var-ious industrial fields. In the formulation of surface coatings and paints, these polymers impart chemical stability, adhesion, mechani-cal properties and optical clarity [40]. The photo-stability of the ali-phatic methacrylic and acrylic polymers is usually higher, when com-pared to polyolefins. Carbonyl ester groups present in these poly-mers are not precisely photo-chemically active. Further, the content of the trace impurities that can initiate the photo-degradation is gen-erally low [41]. On the other hand, it was also proved that the photo-induced oxidation of the methacrylic and acrylic polymers is not auto-catalytical, as seen in the case of polyolefins, however, it advances at a constant oxygen consumption rate [42].

Several studies have reported the weathering degradation behav-ior of the acrylic coatings [1,43]. Hu *et al.* [1] examined the ageing characteristics of the acrylic PU varnish coatings in two artificial weathering environments. The results confirmed that the Xenotest protocol had a remarkable influence on the gloss loss and thickness loss than the UV/condensation weathering exposure. From the FTIR analysis, it was demonstrated that the same degradation mechanism persisted in both weathering conditions.

The thermoset acrylic-melamine coatings are extensively utilized for exterior applications. These are generated by the reaction of acrylic polyol with alkylated melamine formaldehyde (MF) resin. Due to the fact that the reactions are reversible, the acrylic-melamine coatings are prone to degradation when exposed to a weathering environment. From the accelerated tests and outdoor exposures, it has been confirmed that the degradation of these coatings is highly affected by UV radiation, air pollutants and relative humidity [44,45]. Under the influence of UV light, these coatings go through chain scission reactions, thereby, generating amines as well as different carbonyl derivatives as degradation products [46,47].

Nguyen *et al.* [48] studied the effect of relative humidity (RH), ranged from 0-90%, on the moisture enhanced photolysis (MEP) of the moderately methylated melamine acrylic polymer coatings, which were exposed to UV light at 50 °C. Entire degradation under UV conditions at a specified relative humidity could be classified into four modes: the reactions occurring during the post-curing, dark hydrolysis at a specific RH, photolysis and MEP. It was noted that on increasing the RH, the rate and intensity of MEP also enhanced. Further, the increased degradation was described by a mechanism based on the hydrolysis-developed formaldehyde molecules which acted as chromophores for absorbing the UV light, thus, accelerating the photo-oxidation process.

During the photolysis as well as photo-oxidation of poly(methyl methacrylate) (PMMA) at 254 nm, comprehensive random chain breaking, followed by the generation of low molecular weight gas products and monomer molecules were observed [49,50]. The extent of scission was observed to be proportional to the UV radiation dose, and it was greater in inert conditions than air. The scission rate changed with the wavelength of radiation, attaining an ultimate value at 280 nm and reaching zero at wavelengths greater than 320 nm [51]. This phenomenon was described by assuming the effect of ketone or aldehyde groups, which are UV active at 280 nm. The PMMA photolysis occurred as a result of UV absorption, which promoted the de-esterification process due to hemolytic bond breaking. Chiantore *et al.* [52] also examined the photo-oxidative stability of the methacrylic as well as acrylic based polymers, with potential use as protective coatings for various substrates, under artificial solar light radiation. The polymers analyzed in the study included poly(ethyl acrylate), poly(methyl acrylate) (PMA), poly(butyl methacrylate) (PBMA) and poly(ethyl methacrylate). The methacrylate units were observed to be less reactive towards the oxidation process, when compared to the acrylate units. It was also noticed that the degradation of PBMA advanced through an entirely different mechanism, with considerable crosslinking as well as concurrent fragmentation reactions.

Decker *et al.* [53] achieved photo-degradation resistant materials by conserving the surface by using UV cured coatings with HALS radical scavenger as well as a UV absorber (phenyltriazine). The solvent-free UV curable polyurethane-acrylate (PUA) resin consisted of three components, namely aliphatic PU-acrylate telechelic oligomer, hexanediol diacrylate (reactive diluent) and bisacylphosphine oxide photo-initiator. Decker *et al.* [54] also studied the light stability of the water-based UV cured PUA coatings, tested in an accelerated QUV-A weatherometer. The IR spectroscopy was employed to analyze the fast polymerization of the acrylate double bonds at the time of severe illumination. UV curing process was scarcely altered by the incorporation of HALS radical scavengers as well as UV absorbers. The urethane unit (C-NH) was noted to be more sensitive to the photo-degradation process.

In another study, Larche *et al.* [43] predicted the service life of acrylic-melamine and acrylic-urethane coatings under UV light exposure. It was confirmed that the UV irradiation resulted in chain scission and crosslinking in the stabilized and un-stabilized coatings. The breaking of the ether and urethane bonds by UV light caused the generation of free radicals which acted as the source for crosslinking. In the presence of moisture, the photo-degradation of the acrylic-melamine coatings was partially enhanced. The presence of water increased the degradation of moderately methylated melamine acrylic coatings with the generation of formaldehyde by the hydrolysis process. The formaldehyde molecules absorbed UV radiation and dissociated to produce free radicals, followed by the abstraction of hydrogen in the melamine chains, thereby leading to the generation of amine as well as amide based products [48,55]. Also, the weathering degradation of UV cured acrylic coatings constituted chain scission as well as crosslinking. Nevertheless, because of the elaborate network structure, these polymeric coatings were observed to have superior weather resistant than the acrylic-melamine and acrylic-urethane coatings [52-54].

2.3.3 Epoxy Coatings

Epoxies represent an important class of protective polymeric coatings owing to adhesion as well as resistance to chemicals, corrosion, acids and hydrocarbons [56]. During the course of service of the epoxy coatings, the environmental factors like sunlight, atmosphere, temperature and humidity can cause degradation in their properties, thus, limiting their performance [57]. The discoloration and chalking of the epoxy coatings in the presence of UV is regarded as the dominant cause of concern. Several studies have examined the effect of UV light on the degradation behavior of amine cured diglycidyl ether of bisphenol A (DGEBA) epoxy resins [58,59]. It was proposed that the degradation in the chemical structure primarily resulted from the generation of carbonyl products obtained from phenoxy groups, development of amide functions linked to amine concentration and chain breaking processes. Rivaton *et al.* [60] reported that the photo-oxidation process of the phenoxy resins primarily involved the reactions of the aromatic ether units and breaking of CH_3-C bond of the iso-propylidene groups. It was proved that the UV light at the surface of the coating produced a strong oxidative and thermal degradation in the presence of oxygen [61]. The oxidative degradation on the surface of the epoxy coatings activated the microscopic physical defects as well as deformation, present at the molecular and atomic levels at the time of early ageing stages. With increasing the time of irradiation, these physical defects developed and eventually caused the failure of the coating system. Hence, it is critical to understand the changes in the microstructure (like pores, defects and free volume), specifically their impact on the water transportation behavior and anti-corrosion capability of the coatings, during the degradation process.

Fuwei *et al.* [62] examined the development of chemical functional groups, water barrier behavior and microstructure of polyamide-cured epoxy (DGEBA) coatings during UVA photo-oxidative ageing. During the early stages of ageing, reduction in the S parameter as well as the water uptake coefficient illustrated the generation of a highly compact structure caused by post-curing process (S parameter is described as the ratio of the central area to the total area in a Doppler broadening energy spectroscopy (DBES) spectrum). In the subsequent ageing process, the generation of carbonyl groups and molecular chain breaking was observed. From the electrochemical impedance spectroscopy (EIS) analysis, after 208 hours of UV irradiation, a new time constant was observed to develop at relatively high frequency (3.5×10^2 Hz). This revealed that a micro-porous layer developed near the DGEBA film surface.

Using the FTIR analysis, it is possible to predict the long term performance of the materials [63-65]. In one such study, Gerlock *et al.* [66] employed the transmission FTIR analysis and hydroperoxide concentration behavior analysis for comparing the photo-oxidation protection of acrylic-melamine based clearcoats. Penon *et al.* [67] also analyzed the UV degradation behavior of pigmented epoxy coatings for different periods at elevated pressures (1-100 bar). Further, the dielectric sorption analysis (DSA) of the specimens was also carried out. The variation in the dynamics of the absorption characteristics of the degraded polymer resulted due to the enhanced hydrophilicity, porosity and crosslinking. The degradation at all pressures exhibited the desorption behavior, induced by the shrinking of the pore size and swelling of the polymeric coatings. With enhancing the pressure, an increase in the water sorption characteristics was observed in DSA, where a linear trend was observed till 50 bar.

2.3.4 Polyester Coatings

Polyester based coatings are superior candidates for the outdoor applications due to improved mechanical properties as well as outdoor durability. Santos *et al.* [68] examined the degradation performance of polyester and silicone polyester coatings, which were exposed to high UV conditions (two accelerated UV analyses and one natural atmosphere test). The analysis of the coating degradation was performed in accordance with ISO 4628 standard [69]. The coatings exhibited greater color variation and higher gloss loss after exposure of 24 days. It was observed that the organic pigments offered bright colors as well as higher gloss range, however, the inorganic pigments provided better resistance to UV. The polyesters based on isophthalic acid (IPA) have also been generally acknowledged for their weathering stability [70,71].

Adema *et al.* [72] also analyzed the artificial weathering of polyester-urethane coatings using FTIR spectroscopy and UV-vis spectroscopy. The reduction in the urethane crosslinks present in the polymer coatings was observed to take place faster and to a greater extent, when compared to the ester bond breaking. The results obtained from the chemical and optical characterization were used by the authors to propose a kinetic model for the ester bond photolysis, which contributed towards the assessment of the quantum effectiveness of the process.

2.3.5 Silicone Coatings

Silicones, generally termed as polysiloxanes, exhibit high UV resistance because of higher energy bonds and absence of conjugation. Mitra *et al.* [73] examined various dynamic mechanical and physical properties of aliphatic PU, alkyd modified PU, high performance aliphatic PU and cycloaliphatic epoxy modified polysiloxane based coatings before and after artificial weathering. Accelerated weathering test was carried out in a QUV chamber as per ASTM G 154 standard for about 30 days. During the artificial weathering test, the topcoat films were exposed cyclically to 313 nm UV-B radiation at 60 °C for 4 hours, followed by water condensation for 4 hours at 50 °C. Cyclo-aliphatic epoxy modified polysiloxane based coatings displayed nearly no loss of gloss. Overall, these coatings exhibited excellent weatherability as well as good chemical resistance due to comparatively inert Si-O backbone.

2.4 Summary and Outlook

The exposure to ultraviolet (UV) light brings about remarkable degradation in the polymeric coatings. The UV light generates photo-oxidative degradation processes, which cause scission of the polymer chains, thereby producing free radicals and lowering the molecular weight. This induces deterioration in the mechanical properties of the polymeric coatings. The degradation of the polymeric coatings is established in the form of dissolution, crosslinking, color variation, water absorption, oxidation and swelling. The service life of the polymeric coatings can be considerably enhanced, either by the surface treatment to screen the adverse UV radiation or by the incorporation of light stabilizers such as HALS radical scavengers, UV absorbers, etc.

References

1. Hu, J., Li, X., Gao, J., and Zhao, Q. (2009) Ageing behavior of acrylic

polyurethane varnish coating in artificial weathering environments. *Progress in Organic Coatings*, **65**(4), 504-509.

2. Guillet, J. E. (1972) Fundamental Processes in the UV degradation and stabilization of Polymers. *Pure and Applied Chemistry*, **30**(1-2), 135-144.

3. Johnson, B. W., and McIntyre, R. (1996) Analysis of test methods for UV durability predictions of polymer coatings. *Progress in Organic Coatings*, **27**(1), 95-106.

4. Yang, X. F., Tallman, D. E., Bierwagen, G. P., Croll, S. G., and Rohlik, S. (2002) Blistering and degradation of polyurethane coatings under different accelerated weathering tests. *Polymer Degradation and Stability*, **77**(1), 103-109.

5. *Handbook of UV Degradation and Stabilization*, Wypych, G. (ed.), ChemTec Publishing, USA (2010).

6. *Handbook of Polymer Degradation*, Hamid, S. H., Amin, M. B., and Maadhah, A. G. (eds.), Marcel Dekker, USA (1992).

7. Katangur, P., Patra, P. K., and Warner S. B. (2006) Nanostructured ultraviolet resistant polymer coatings. *Polymer Degradation and Stability*, **91**(10), 2437-2442.

8. Revie, R. W. (2008) *Corrosion and Corrosion Control: An Introduction to Corrosion Science and Engineering*, 4th edition, John Wiley & Sons, USA.

9. Ahmad, Z. (2006) Principles of Corrosion Engineering and Corrosion Control, Butterworth-Heinemann, UK.

10. Gardette, J.-L., Colin, A., Trivis, S., German, S., and Therias, S. (2014) Impact of photooxidative degradation on the oxygen permeability of poly(ethyleneterephthalate). *Polymer Degradation and Stability*, **103**, 35-41.

11. Croll, S. G., and Skaja, A. D. (2003) Quantitative spectroscopy to determine the effects of photodegradation on a model polyester-urethane coating. *Journal of Coatings Technology*, **75**, 85-93.

12. Skaja, A. D., and Croll, S. G. (2003) Quantitative ultraviolet spectroscopy in weathering of a model polyester-urethane coating. *Polymer Degradation and Stability*, **79**, 123-131.

13. Nichols, M., Boisseau, J., Pattison, L., Campbell, D., Quill, J., Zhang, J., Smith, D., Henderson, K., Seebergh, J., Berry, D., Misovski, T., and Peters, C. (2013) An improved accelerated weathering protocol to anticipate Florida exposure behavior of coatings. *Journal of Coatings Technology and Research*, **10**, 153-173.

14. Glockner, P., Ritter, H., Osterhold, M., Buhk, M., and Schlesing, W. (1999) Effect of weathering on physical properties of clearcoats. *Die Angewandte Makromolekulare Chemie*, **269**, 71-77.

15. Croll, S. G., Hinderliter, B. R., and Liu, S. (2006) Statistical approaches for predicting weathering degradation and service life. *Progress in Organic Coatings*, **55**, 75-87.

16. Nichols, M. E., Gerlock, J. L., Smith, C. A., and Darr, C. A. (1999) The effects of weathering on the mechanical performance of automotive paint systems. *Progress in Organic Coatings*, **35**, 153-159.
17. Larché, J. F., Bussière, P. O., Wong-Wah-Chung, P., and Gardette, J. L. (2012) Chemical structure evolution of acrylic-melamine thermoset upon photo-ageing. *European Polymer Journal*, **48**, 172-182.
18. Croll, S. G., Shi, X., and Fernando, B. M. D. (2008) The interplay of physical aging and degradation during weathering for two cross-linked coatings. *Progress in Organic Coatings*, **61**, 136-144.
19. Gu, X., Michaels, C. A., Drzal, P. L., Jasmin, J., Martin, D., Nguyen, T., and Martin, J. W. (2007) Probing photodegradation beneath the surface: a depth profiling study of UV-degraded polymeric coatings with microchemical imaging and nanoindentation. *Journal of Coatings Technology and Research*, **4**, 389-399.
20. Larche, J. F., Bussiere, P. O., and Gardette, J. L. (2011) Photo-oxidation of acrylic-urethane thermoset networks. Relating materials properties to changes of chemical structure. *Polymer Degradation and Stability*, **96**, 1438-1444.
21. Croll, S. G. (2013) Application and limitations of current understanding to model failure modes in coatings. *Journal of Coatings Technology and Research*, **10**, 15-27.
22. Decker, C., and Zahouily, K. (1999) Photodegradation and photooxidation of thermoset and UV-cured acrylate polymers. *Polymer Degradation and Stability*, **64**, 293-304.
23. He, Y., Yuan, J. P., Cao, H., Zhang, R., Jean, Y. C., and Sandreczki, T. C. (2001)Characterization of photo-degradation of a polyurethane coating system by electron spin resonance, *Progress in Organic Coatings*, **42**(1-2), 75-81.
24. *Characterization of Polymers*, Tong, H.-M., Kowalczyk, S. P., Saraf, R., and Chou N. J. (eds.), Butterworths Heinemann, USA (1994).
25. *Multidimensional Spectroscopy of Polymers: Vibrational, NMR, and Fluorescence Techniques*, Urban, M. W., and Provder, T. (eds.), American Chemical Society, USA (1995).
26. Cao, H., Zhang, R., Sundar, C. S., Yuan, J.-P., He, Y., Sandreczki, T. C., Jean, Y. C., and Nielsen, B. (1998) Degradation of polymer coating systems studied by positron annihilation spectroscopy. 1. UV irradiation effect. *Macromolecules*, **31**(19), 6627-6635.
27. Allen, N. S., and Edge, M. (1993) *Fundamentals of Polymer Degradation and Stabilization*, Springer, Netherlands.
28. Asmatulu, R., Claus, R. O., Mecham, J. B., and Corcoran, S. G. (2007) Nanotechnology-associated coatings for aircrafts, *Journal of Materials Science*, **43**, 415-422.
29. Fernando, R. H. (2009) Nanocomposite and nanostructured coatings: recent advancements, Nanotechnology Applications in Coating. *ACS Symposium Series*, **1008**, 2-21.

30. Asmatulu, R., and Revuri, S. (2008) Synthesis and Characterization of Nanocomposite Coatings for the Prevention of Metal Surfaces. *Proceedings of SAMPE Fall Technical Conference*, USA, pp. 1-13.
31. Adema, K. N. S. (2015) *Photodegradation of Polyester-Urethane Coatings*, Technische Universiteit Eindhoven, Netherlands.
32. Allen, N. S., Chirinis-Padron, A., and Henman, T. J. (1985) The Photo-stabilisation of Polypropylene: A Review. *Polymer Degradation and Stability*, **13**, 31-76.
33. White, J. R., and Turnbull, A. (1994) Weathering of polymers: mechanisms of degradation and stabilization, testing strategies and modelling. *Journal of Materials Science*, **29**, 584-613.
34. *Polyurethane Handbook*, Oertel, G. (ed.), Hanser Publishers, Germany (1985).
35. Roffey, C. G. (1982) *Photopolymerization of Surface Coatings*, John Wiley & Sons, USA.
36. Rabek, J. F. (1990) *Photostabilization of Polymers: Principles and Applications*, Elsevier Applied Science, UK.
37. Kachan, A. A., Kargan, N. P., Kulik, N. V., and Boyarskii, G. Y. (2004) Two-photon heterogeneous photodegradation of an aromatic polyurethane. *Theoretical and Experimental Chemistry*, **4**, 314-317.
38. Singh, R. P., Tomer, N. S., and Bhadraiah, S. V. (2001) Photo-oxidation studies on polyurethane coating: effect of additives on yellowing of polyurethane. *Polymer Degradation and Stability*, **73**(3), 443-446.
39. Decker, C., and Bandaikha, T. (1989) *International Conference on Advances in the Stabilization and Controlled Degradation of Polymers*, Technomic Publishing, UK, p. 143.
40. *Surface Coatings: Science and Technology*, Paul, S. (ed.), 2nd edition, Wiley, UK (1995).
41. Davis, A., and Sims, D. (1983) *Weathering of Polymers*, Elsevier, Netherlands.
42. Carduner, K. R., CarterIII, R. O., Zimbo, M., Gerlock, J. L., and Bauer, D. R. (1988) End groups in acrylic copolymers. 1. Identification of end groups by carbon-13 NMR. *Macromolecules*, **21**(6), 1598-1603.
43. Larche, J. F., Bussiere, P. O., and Gardette, J. L. (2010) How to reveal latent degradation of coatings provoked by UV-light, *Polymer Degradation and Stability*, **95**(9), 1810-1817.
44. *Characterization of Highly Crosslinked Polymers*, Labana, S. S., and Dickie, R. A. (eds.), American Chemical Society, USA (1983).
45. Schmitz, P. J., Holubka, J. W., and Xu, L. F. (2000) Mechanism for environmental etch of acrylic melamine-based automotive clearcoats: Identification of degradation products. *Journal of Coatings Technology*, **72**(904), 39-45.
46. *Service Life Prediction of Organic Coatings: A Systematic Approach*, Bauer, D., and Martin, J. W. (eds.), American Chemical Society, USA

(1999).

47. Bauer, D. R., and Mielewski, D. F. (1993) The role of humidity in the photooxidation of acrylic melamine coatings. *Polymer Degradation and Stability*, **40**, 349-355.

48. Nguyen, T., Martin, J., Byrd, E., and Embree, N. (2002) Relating laboratory and outdoor exposure of coatings III. Effect of relative humidity on moisture-enhanced photolysis of acrylic-melamine coatings. *Polymer Degradation and Stability*, **77**, 1-16.

49. Shultz, A. R. (1961) Degradation of polymethyl methacrylate by ultraviolet light. *The Journal of Physical Chemistry*, **65**, 967-972.

50. Allison, J. P. (1966) Photodegradation of poly (methyl methacrylate). *Journal of Polymer Science, Part A: Polymer Chemistry*, **4**(5), 1209-1221.

51. Torikai, A. Ohno, M., and Fueki, K. (1990) Photodegradation of poly (methyl methacrylate) by monochromatic light: Quantum yield, effect of wavelengths, and light intensity. *Journal of Applied Polymer Science*, **41**, 1023-1032.

52. Chiantore, O., Trossarelli, L., and Lazzari, M. (2000) Photooxidative degradation of acrylic and methacrylic polymers, *Polymer*, **41**, 1657-1668.

53. Decker, C., and Zahouily, K. (2002) Photostabilization of polymeric materials by photoset acrylate coatings. *Radiation Physics and Chemistry*, **63**, 3-8.

54. Decker, C. Masson, F., and Schwalm, R. (2004) Weathering resistance of water based UV-cured polyurethane-acrylate coatings. *Polymer Degradation and Stability*, **83**, 309-320.

55. Nguyen, T., Martin, J., Byrd, E., and Embree, N. (2002) Relating laboratory and outdoor exposure of coatings: II. Effects of relative humidity on photodegradation and the apparent quantum yield of acrylic-melamine coatings. *Journal of Coatings Technology*, **74**, 65-80.

56. Malshe, V. C., and Waghoo, G. (2004) Weathering study of epoxy paints. *Progress in Organic Coatings*, **51**, 267-272.

57. Huang, W., Zhang, Y., Yu, Y., and Yuan, Y. (2007) Studies on UV-stable silicone-epoxy resins. *Journal of Applied Polymer Science*, **104**, 3954-3959.

58. Bellenger, V., and Verdu, J. (1983) Photooxidation of amine cross-linked epoxies I. The DGEBA-DDM system. *Journal of Applied Polymer Science*, **28**, 2599-2609.

59. Kim, H., and Urban, M. W. (2000) Molecular level chain scission mechanisms of epoxy and urethane polymeric films exposed to UV/H_2O. Multidimensional spectroscopic studies, *Langmuir*, 16, 5382-5390.

60. Rivaton, A. Moreau, L., and Gardette, J-L. (1997) Photo-oxidation of phenoxy resins at long and short wavelengths-I. Identification of the

photoproducts. *Polymer Degradation and Stability*, **58**, 321-331.

61. Mailhot, B., Morlat-Therias, S., Bussiere, P.-O., and Gardette, J-L. (2005) Study of the degradation of an epoxy/amine resin, 2. *Macromolecular Chemistry and Physics*, **206**(5), 585-591.

62. Liu, F., Yin, M., Xiong, B., Zheng, F., Mao, W., Chen, Z., He, C., Zhao, X., and Fang, P. (2014) Evolution of microstructure of epoxy coating during UV degradation progress studied by slow positron annihilation spectroscopy and electrochemical impedance spectroscopy. *Electrochimica Acta*, **133**, 283-293.

63. Rabek, J. F. (1995) *Polymer Photodegradation: Mechanisms and Experimental Methods*, Springer, Netherlands.

64. Bellinger, V., Bouchard, C., Claveirolle, P. and Verdu, J., (1981) Photo-oxidation of epoxy resins cured by non-aromatic amines. *Polymer Photochemistry*, **1**(1), 69-80.

65. Bellinger, V., and Verdu, J. (1985) Oxidative Skeleton Breaking in Epoxy-Amine Networks. *Journal of Applied Polymer Science*, **30**(1), 363-374.

66. Gerlock, J. L., Smith, C. A., Nunez, E. M., Cooper, V. A., Liscombe, P., Cummings, D. R., and Dusibiber, T. G., (1996) Measurements of chemical change rates to select superior automotive clearcoats. *Advances in Chemistry*, **249**, 335-347.

67. Penon, M. G., Picken, S. J., Wűbbenhorst, M., and van Turnhout, J. (2007) Dielectric sorption analysis of pigmented epoxy coatings UV degraded at elevated pressures. *Polymer Degradation and Stability*, **92**(10), 1857-1866.

68. Santos, D., Costa, M. R., and Santos, M. T. (2007) Performance of polyester and modified polyester coil coatings exposed in different environments with high UV radiation. *Progress in Organic Coatings*, **58**(4), 296-302.

69. *ISO 4628, Paints and Varnishes: Evaluation of Degradation of Coatings. Designation of Quantity and Size of Defects, and Intensity of Uniform Changes in Appearance*, International Organization for Standardization (2016). Online: https://www.iso.org/standard/64877.html [accessed 21st March 2019].

70. Maetens, D. (2007) Weathering degradation mechanism in polyester powder coatings. *Progress in Organic Coatings*, **58**, 172-179.

71. Molhoek, L., Posthuma, C., and Gijsman, P. (2013) Weathering well. *European Coatings Journal*, 79-82.

72. Adema, K. N. S., Makki, H., Peters, E. A. J. F., Laven, J., van der Ven, L. G. J., van Benthem, R. A. T. M., and de With, G. (2014) Depth-resolved infrared microscopy and UV-VIS spectroscopy analysis of an artificially degraded polyester-urethane clearcoat. *Polymer Degradation and Stability*, **110**, 422-434.

73. Mitra, S., Ahire, A., and Mallik, B. P. (2014) Investigation of accelerated aging behaviour of high performance industrial coatings by

dynamic mechanical analysis. *Progress in Organic Coatings*, **77**(11), 1816-1825.

3

Biodegradation Properties of Melt Processed PBS/Chitosan Bio-nanocomposites with Silica, Silicate and Thermally Reduced Graphene

3.1 Introduction

Majority of traditional polymers are non-biodegradable and their recycling or re-use is very challenging, thus, contributing to piles of non-biodegradable wastes all over the world. Due to increased concerns about the environment and pollution hazards, the use of biodegradable polymers and their composites has gained interest in the recent years [1]. For instance, the introduction of green bags in the Australian markets has resulted in polyethylene (PE) bag usage to fall from around 6 billion in 2002 to 3.9 billion in 2007 [2]. In Europe, despite the economic crisis, the market of biodegradable polymers grew 5-10% (depending on products and applications) in 2009 compared with 2008. A report from Business Communications Company (BCC) Research estimated a compound annual growth rate (CAGR) of 22% for the biodegradable polymers for the 5-year period starting from 2012 [3].

A biodegradable plastic is generally defined as a plastic in which the degradation results from the action of naturally occurring micro-organisms such as bacteria, fungi and algae. Likewise, a compostable plastic is classified as a plastic which undergoes degradation by the biological processes to generate CO_2, water, inorganic compounds and biomass without leaving any toxic residue [4]. Several studies have been performed to gain insights about the mechanism and estimation of polymer biodegradation [5-8]. Important factors affecting the biodegradability are the concentration of the test material, physico-chemical properties, inorganic nutrients in the test medium, presence of other degradable substances and test

Fakhruddin Patwary[a], Nadejda Matsko[b] and Vikas Mittal[a],*
[a]The Petroleum Institute (part of Khailfa University of Science and Technology), Abu Dhabi, UAE; [b]Graz Centre for Electron Microscopy, Graz, Austria
*Current address: Bletchington, Wellington County, Australia

conditions (duration, temperature, open/closed vessels, size of vessels, etc.).

Major drawbacks of the biodegradable polymers hindering their widespread application are their high cost and poor mechanical properties as compared to the conventional thermoplastics [9]. In recent years, the inclusion of the nanoparticles into these biodegradable polymer matrices has attracted a large research effort in order to develop polymeric materials with improved/desired properties with varying degree of success [10]. Though the reported studies focused on the enhancement of the mechanical properties of the biopolymers on incorporation of fillers, the effect of fillers on the biodegradability of the polymer matrix in the composite has largely been ignored.

Chitosan (CS) is a biopolymer derived by deacetylation of chitin and is non-toxic, biodegradable and second most abundant biopolymer in nature after cellulose [11]. Various crosslinking agents like glutaraldehyde, sulfuric acid, sulfonic acid groups, etc., have been reported to enhance its mechanical properties [12], however, limited research has been performed on the melt processing of the chitosan based materials. Chitosan based composites have great potential for application in membrane technology [12,13], water treatment [14], catalysis [15,16], biomedical appliances [17,18], etc. There have also been many studies reporting the blends and composites based on chitosan and biodegradable aliphatic polyesters like poly(butylene succinate) (PBS) [19-27], but no insights into the biodegradability of these composite systems have been provided. PBS is a biodegradable aliphatic polyester derived from the petrochemical sources and has highly crystalline morphology with ease of processing [28], thus, the blends and composites based on PBS have high commercial potential.

Some studies on the soil burial tests as a mean for analyzing the biodegradability of composites have been reported [29-34]. Wu [29] reported polycaprolactone (PCL)/chitosan and acrylic acid grafted polycaprolactone (PCL-g-AA)/chitosan composites. It was found that for both PCL/chitosan and PCL-g-AA/chitosan, the weight loss was high for the samples with higher content of chitosan. Kim *et al.* [32] analyzed the biodegradability of bio-flour filled PBS composites in both natural and aerobic compost soil. As expected, the biodegradability of these composites increased with increasing the content of biodegradable bio-flour, when aerobic composting soil was used [32]. Several test methods and standards for testing the biodegradability of polymers like CO_2 evolution, aerobic composting and respirometric tests were also discussed by Pagga [5], and it was suggested that the aerobic composting test (compost as inoculum, 60 °C, CO_2 measurement) should be used for the biodegradation analysis of the polymeric materials.

In the current study, the biodegradation analysis of the bio-nanocomposites utilizing a blend of PBS and chitosan with silica, silicate and graphene reinforcements has been performed. The mechanical properties of these composites were reported elsewhere [21] and the high extent of property enhancements confirmed the generation of functional materials capable of replacing non-biodegradable materials in different commercial applications. Silica, layered silicates as well as graphene form important categories of 0-D and 2-D nano-particulate reinforcements which lead to significant enhancements in the mechanical and thermal properties of the polymers [23,35-39], however, the impact on the polymer biodegradability on incorporation of these fillers is required to be systematically studied. In this study, the biodegradability of these bio-nanocomposites has been studied using the soil burial test under natural conditions followed by extensive thermal and morphological characterization.

3.2 Experimental

3.2.1 Materials

Chitosan flakes, procured from Bio21, Thailand, were donated by Prof. Wang at The Petroleum Institute. Chitosan had a degree of deacetylation (DD) of 90% and weight average molecular weight (M_w) of 180000 g/mol. PBS (Bionelle 1050 from Showa Highpolymer Co., Japan) was kindly donated by Prof. Shih from Chaoyang University, Taiwan. Synthetic silicon dioxide powder (ZEOFREE® 5161 S) and synthetic alumino-silicate (ZEOLEX® 23) were supplied by J. M. Huber Private Limited, India. The polymer and filler materials were used as received. Thermally reduced graphene was produced by thermal exfoliation of the graphite oxide precursor using modified Hummer's method, as reported earlier [38,40].

3.2.2 Preparation of Nanocomposites

HAAKE Minilab micro-compounder from Thermo Scientific was employed for the melt mixing process. It had a conical twin-screw geometry and counter rotating mode was used for extrusion. The filler and polymer materials were physically premixed in a small bowl before feeding into the extruder. To generate the nanocomposites, the compounding temperature of 170 °C, screw speed of 100 rpm and mixing time of 15 min were used. PBS:CS (60:40) nanocomposites with filler contents ranging from 2 wt% to 10 wt% were generated. In addition, the PBS:CS (60:40) blend was also melt processed similarly. Tetrahedron MTP-10 hot press was used to compression mold the test samples. Hot pressing was performed at 170 °C with a pressure of 800 bar and a holding pressure of 290 bar. The samples were subsequently cooled down to 30 °C at a rate of 5 °C/min. Sheets of 12 x 12 cm with a thickness of 1.5 mm were produced and test samples of desired shape were stamped out from these sheets.

3.2.3 Soil Burial Test

Biodegradability of PBS:CS blend based nanocomposites was analyzed with soil burial tests. The tests were carried out under natural environmental conditions and no additional effort was made to modify the conditions. The samples were buried in commercially available compost soil for gardening (Bestgreen Universal potting soil). Two pots of 52 x 14 x 10 cm dimensions were used for the soil burial test and placed in an open place usually used for gardening. Soil was filled to a height of 10 cm in the pots. Samples with dimension of 17 x 9.5 x 1.5 mm were buried 5 cm deep, 4 cm apart in the longitudinal direction and 5 cm apart in the transverse direction. The soil moisture content was 60 to 70% on a normal day. On a sunny day, the soil temperature was usually between 33 to 37 °C at mid-day and between 20 to 25 °C at night. The samples were dug out every 15 days (d), cleaned from soil particles and then dried in a vacuum oven at 70 °C. Their weight was measured at regular intervals until the weight became constant.

3.2.4 Characterization Techniques

Differential scanning calorimetric (DSC) measurements on the polymer blend and bio-nanocomposites were performed using 204 F1 Phoenix Netzsch DSC under nitrogen atmosphere. The scans were obtained from 30-200-30 °C using the heating and cooling rates of 5

°C/min respectively. Thermal properties of the samples were analyzed using Netzsch STA 409 PC Luxx thermogravimetric analyzer
(TGA). Nitrogen was used as a carrier gas and the scans were obtained from 50 to 700 °C at a heating rate of 20 °C/min.

LINSEIS STA PT1600 TGA system was also used to measure the
weight change during heating in air atmosphere. This system was
coupled with a Phipher mass spectrometer (MS) which allowed the
determination of the elimination of H_2O (m/z=18), CO_2 (m/z=44)
and other fractions during heating. All measurements were realized
with a heating rate of 3 °C/min in the temperature range of 25 to
800 °C, using a flow rate of 20 ml/min.

Infra-red (IR) spectroscopic analysis was performed on a Nicolet
iS10 spectrometer equipped with SmartiTR diamond ATR accessory
(angle of incidence of 45°), DTGS KBr detector and KBr beam split-
ter. It had a diamond ATR crystal (index of refraction 2.4 at 1000
cm^{-1}) and a depth penetration of 2 μm at 1000 cm^{-1} for a sample
with refractive index of 1.5. Spectra were recorded by the OMNIC
software in the 4000-525 cm^{-1} region with a resolution of 4 cm^{-1}
from 32 scans.

Solid-state ^{13}C NMR experiments were conducted on a Bruker
400MHz spectrometer at the resonance frequency of 100.628 MHz
for ^{13}C and 400.133 MHz for ^{1}H using cross- polarization (CP) and
MAS. Around 100 mg of the extruded sample was tightly packed into
a 4 mm diameter cylindrical zirconium oxide rotor with a KelF end
cap. The rotor was spun at 8 kHz at the magic angle 54.7°. ^{13}C
CP/MAS NMR spectra were measured with a CP contact time of 1
ms, recycle delay of 5 s, 3.2 μs 900 pulse and acquisition time of 20
ms. A total of 1024 scans were typically averaged for each sample.
All NMR experiments were carried out at ambient probe temperature (25 °C) with high-power proton decoupling during acquisition.

For the microscopy analysis, the samples were mounted in special holders which could fit in the microtome and were suitable at
the same time for the examination of the block face by atomic force
microscopy. Block faces of the samples were obtained using a Leica
Ultracut E microtome (Leica, Austria) equipped with a diamond
knife (Diatome, Switzerland) at -120 °C. AFM analysis of the block
face of the samples was obtained in tapping mode at ambient conditions using a Digital Instruments NanoScope III and silicon nitride
cantilevers with natural frequencies in the 300 kHz range (force
constant 20 N/m, tip radius 10 nm (NT-MDT, Russia)). The block
faces of the specimens after cryo-ultramicrotomy were also investi-

gated at ambient conditions using ZEISS Axioplan light microscope (ZEISS, Germany) in reflected polarized light. The microscope was equipped with a ZEISS Axio Cam ICc 1 CCD camera.

3.3 Results and Discussion

Figure 3.1 represents the relative mass loss (mass of degraded samples/initial mass) as a function of embedding time for PBS:CS nanocomposites (normalized to actual polymer mass in the composites). The mass of the samples decreased with embedding time confirming the initiation of degradation in the natural composting conditions. PBS:CS blend exhibited the maximum extent of biodegradation in comparison with the nanocomposites. Nearly 4% loss in the sample mass was observed after 60 d of embedding in normal environmental conditions. The reduced extent of mass loss in the nanocomposites was expected as the addition of nanofillers hindered the access of the microbes to the polymer chains. Thus, the microbes had to traverse longer and tortuous pathways in the polymer matrix, which subsequently slowed the biodegradation in the nanocomposites [32]. Though the biodegradation was reduced in the presence of fillers, but it was still appreciable especially in silica and silicate nanocomposites considering that the experiments were performed in natural conditions at room temperature during fall season. Increasing the amount of filler in the silica and silicate composites also increased the amount of weight loss due to a decrease in the overall strength of the polymer on addition of a large amount of filler particles and also due to the resulting filler agglomerates. Also, as reported elsewhere [21], the tensile strength of the polymer matrix decreased from 19 MPa to 13, 14 and 12 MPa for silica, silicate and graphene composites respectively with 10% filler content. As observed in Figure 3.1(a) and (b), the nanocomposites with silica and silicate showed gradual reduction in the weight ratio till 45 d. On the other hand, the behavior of the graphene nanocomposites was different as initial gain in mass was observed in the samples (Figure 3.1(c)). This might have resulted from the presence of functional groups on the surface of polar graphene platelets which led to the absorption and retention of moisture in the graphene composites. Difficulty of penetration faced by microbes would also be specifically severe in this case owing to the high aspect ratio graphene platelets. The rate of degradation was also observed to slow down to a small extent in the 45-60 d period, thus, indicating that a specific

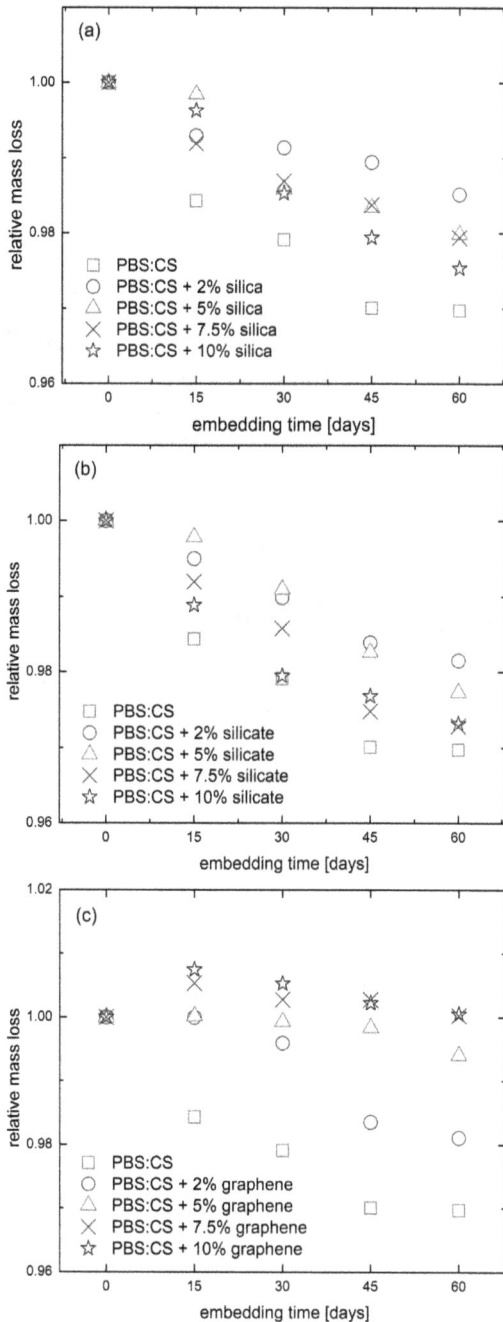

Figure 3.1 Mass loss of the samples as a function of embedding time; (a) silica, (b) silicate and (c) graphene nanocomposites.

phenomenon associated with biodegradation might be taking place in the composites.

Figure 3.2 compares the optical micrographs of the fresh surface of the nanocomposites (5% filler content) with the samples embedded for 30 and 60 d in soil. Before burial (i.e. at day 0), all the samples had relatively smooth and crack-free surface. The surface of the

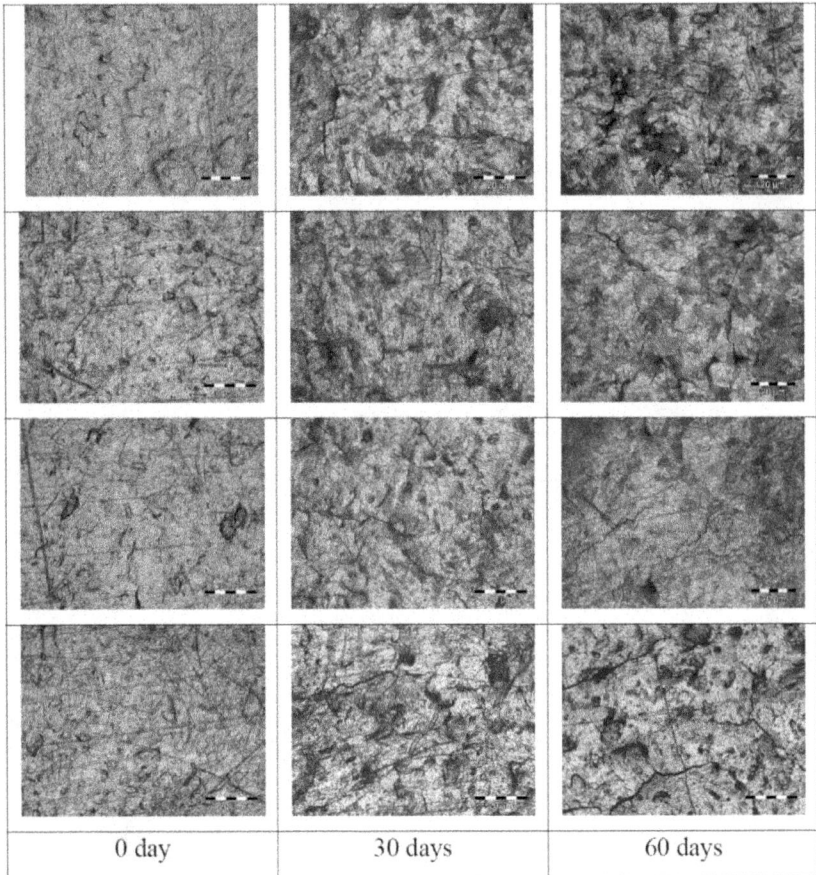

0 day	30 days	60 days

Figure 3.2 Optical micrographs of the surface of the composites as a function of embedding time; (1st row) pure PBS:CS blend, (2nd row) 5% silica composites, (3rd row) 5% silicate composites and (4th row) 5% graphene composites. The scale bar in the images reads 20 μm.

samples was observed to become rough with increasing embedding period and more prominent surface cracks appeared, thus, supporting the loss of mass of the samples due to appreciable biodegradation. Qualitative comparison of the micrographs of the three nano-composites also revealed similar extent of biodegradation in all samples. Thus, though a significant weight loss was not observed at the natural conditions of biodegradation testing, the changes in the microstructure signified that the degradation was initiated, thus, causing an extensive effect on the polymer matrix. Similarly, Figure 3.3(a) also shows the changes on the surface of PBS:CS+10% silicate

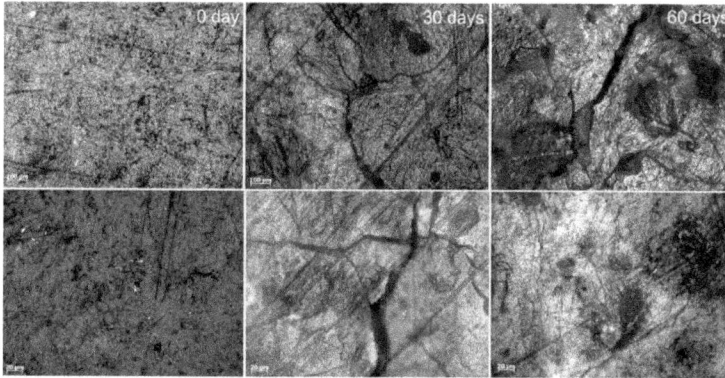

(a)

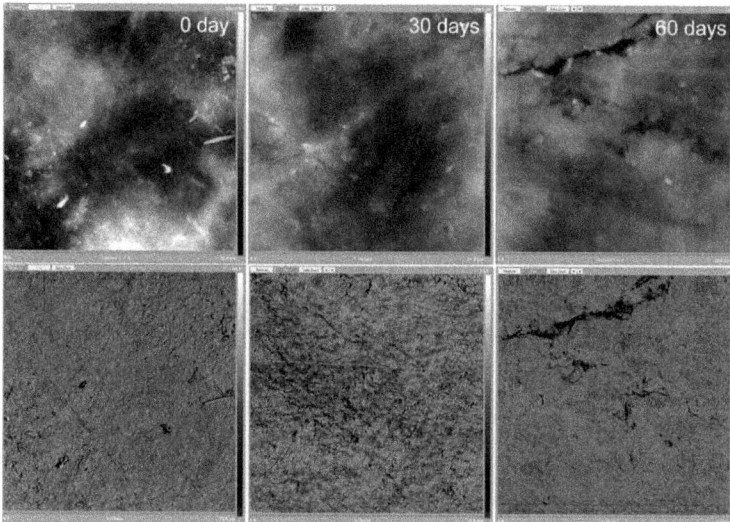

(b)

Figure 3.3 (a) Optical images of the PBS:CS+10% silicate nanocomposite as a function of time; (first column) 0 days, (middle column) 30 days and (third column) 60 days. Figure (b) represents the AFM images of the PBS:CS+10% silica nanocomposite system.

composite as a function of embedding time. In comparison with the composites with 5% filler content, the optical micrographs revealed that the extent of degradation with embedding time was more severe, as also observed in the mass loss analysis. Even after 30 d of embedding, a large network of interconnected cracks of different sizes was observed in the composite, which were totally absent in the sample before embedding. Silica composite with 10% filler content also exhibited similar behavior, as shown in the AFM height and phase images in Figure 3.3(b). The ultrastructure of the polymer was observed to be altered significantly after 30 d of soil embedding and the loss of dimensional stability of the polymer was confirmed by the large number of cracks. The cracks were observed to be widened significantly in the composite after 60 d of embedding. Moreover, some chitosan platelets were observed on the surface of the composite before soil burial, which were completely absent in the degraded samples. It may indicate that chitosan degraded preferentially as compared to PBS in the nanocomposites. In contrast, the composite with 10% graphene content though showed degradation, but its extent was lower than the corresponding silica and silicate nanocomposites due to the earlier mentioned reasons.

As observed in the mass loss and optical microscopy studies, the biodegradation was affected by the type of filler, its content in the composites, its effect on the polymer structure as well as the amount of embedding time. Further insights into the effect of filler addition on the biodegradation of the polymer matrix were obtained from the DSC studies of the crystalline melting of the PBS phase. Chitosan flakes did not show any melt transition in the range used for the DSC analysis and remained solid throughout the melt extrusion process. As observed from the thermograms corresponding to the PBS:CS nanocomposites with 2% filler content (Figure 3.4), the peak melting point (T_m) shifted to higher temperatures for the samples embedded for 15 and 30 d. Subsequently, with increased embedding time of 45 and 60 d, the peak melting point exhibited no change or minor decrease in its value. Similar phenomenon was observed for all filler types and concentrations. The PBS melting transitions in the composites were also broadened after soil embedding, though there was no specific effect of embedding time. This clearly indicated that the composites went through some structural change during increasing period of soil burial. The increase in T_m in the initial degradation phase indicated that the overall crystallinity might have increased, which was also expected due to the fact that the microbes

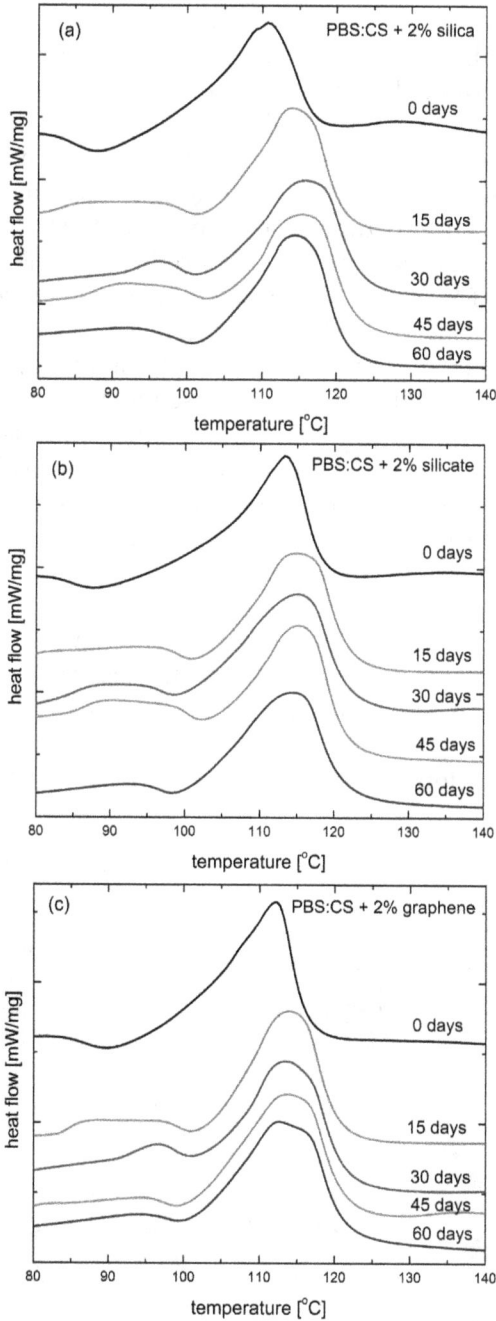

Figure 3.4 DSC melting thermograms of PBS:CS+2% filler composites as a function of embedding time in soil; (a) silica, (b) silicate and (c) graphene.

preferentially degrade the random amorphous component of the polymer, resulting in an increase in the fraction of the crystalline phase [41]. Presumably, by 30 to 45 d, easily accessible amorphous portion of the composites was degraded by the microbes followed by attack on the ordered or crystalline regions which caused the decrease in T_m and crystallinity. Rate of weight loss was also observed to decrease during this time of 45 and 60 d of embedding due to a slower rate of degradation of the ordered regions. From the observations in the AFM images (Figure 3.3(b)) and the increased crystallinity of the PBS phase, this loss of material can also be correlated to the preferential degradation of chitosan in the composites, however, the amorphous content of the PBS phase would also be degraded in the process. Figure 3.5 also shows the cross-section of the PBS:CS+10% silicate nanocomposite degradation as a function of embedding time. Before burial, the chitosan flakes were observed to be uniformly incorporated in the PBS matrix and no interfacial debonding was present. Initiation of degradation was observed after 15 d of embedding and the interfacial areas between the PBS and chitosan phases were affected. After 30 d of embedding, extensive degradation took place as confirmed by the loss of structure, and the chitosan platelets were observed to be largely absent from the cross-section. It confirmed the earlier indications from AFM and DSC that the degradation of chitosan preceded the crystalline phase degradation of PBS.

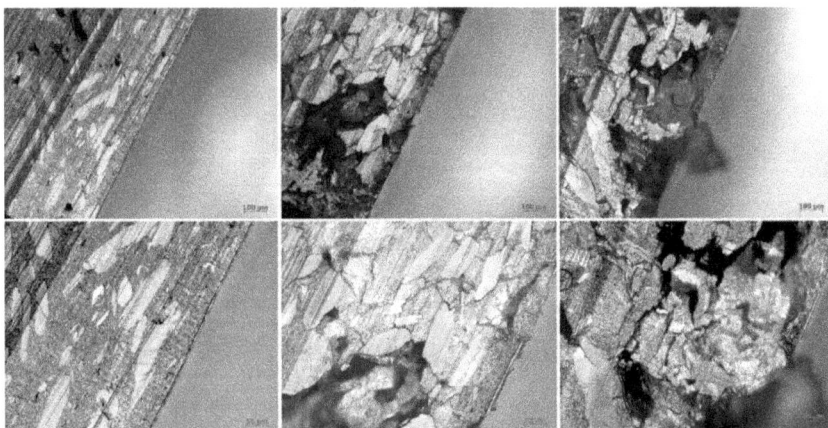

Figure 3.5 Optical images of the PBS:CS+10% silicate nanocomposites showing the cross-section in relation with the embedding time; (first column) 0 days, (middle column) 15 days and (third column) 30 days.

To further confirm the notion of change in crystallinity with embedding, relative percent crystallinity of the composites was plotted in Figure 3.6 as a function of embedding time. Relative percent crystallinity was calculated from the ratio of the enthalpy of the melting transition of the sample at corresponding day of embedding and the melt enthalpy of the sample before embedding. In all samples, the crystallinity was observed to increase till 30 d of embedding followed by minor decrease. PBS:CS blend exhibited significant increase in relative crystallinity due to a higher extent of degradation, as observed earlier. For silica composites, the composites with 2 and 5% filler content had lower extent of crystallinity increase with time, however, the composites with 7.5 and 10% filler had even higher extent of increase in crystallinity than the pure blend. It confirmed that the behavior of nanocomposites towards biodegradation was a compromise of many influencing factors like generation of tortuous path for microbes due to filler addition leading to lower rate of degradation as well as lower polymer strength helping in faster rate of degradation. In silicate composites, the crystallinity increased as a function of filler content, but the values were always smaller than the pure blend. As observed earlier, the graphene composites exhibited smallest increase in crystallinity. The increase was a weak function of filler content, which indicated that the phenomenon of increase in mass observed earlier did not hinder the polymer degradation. Peaking of crystallinity at 30 d of embedding for all samples also confirmed that the mechanism of degradation in the composites was still the same, irrespective of the filler type and amount. Figure 3.7 shows the CP/MAS spectra of PBS:CS+5% graphene composite after 0, 15 and 30 d of embedding. Comparison of the spectra indicated that the chemical shifts of the polymers did not change positions in the nanocomposites with embedding time, but changes in the intensity and broadness were apparent for chemical shifts. For instance, shifts in the region 70-80 ppm, 60-70 ppm and 45-55 ppm became more sharp and intense, thus, confirming that there was an overall decrease in the amorphous content in the nanocomposites with increasing embedding time. Similar observations were noticed for composites containing silica and silicate.

Thermal degradation of the biodegraded composites was studied through TGA in order to monitor the changes in thermal degradation patterns as a function of degradation time under soil. The TGA thermograms of the blend and composites samples in nitrogen atmosphere exhibited two mass loss regions, one exhibiting the peak

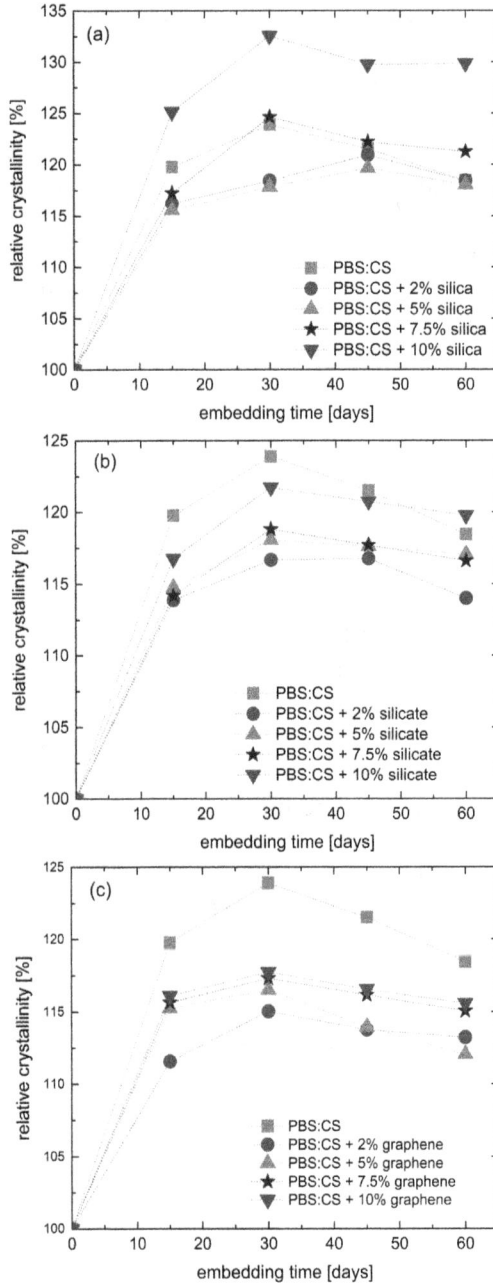

Figure 3.6 Relative crystallinity of the nanocomposites before and after soil burial. Figures (a), (b) and (c) represent silica, silicate and graphene nanocomposite systems respectively.

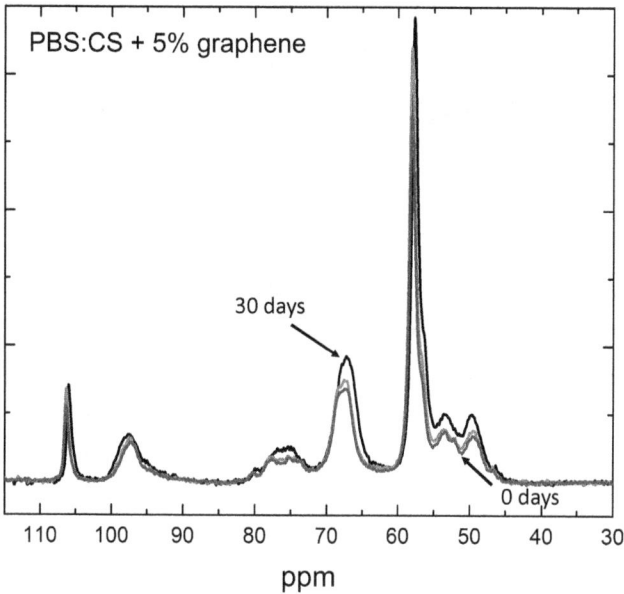

Figure 3.7 Solid state NMR spectra of PBS:CS+5% graphene composites at 0, 30 and 60 days of embedding in soil.

degradation temperature of 310 °C corresponding to chitosan degradation and other with a peak degradation peak around 390 °C due to PBS. These separate peaks due to the chitosan and PBS phases confirmed that the polymers did not mix at molecular level, as observed earlier. In the thermograms of the samples after embedding, no change in the degradation mechanism was observed. However, the chitosan degradation peak was observed to be shifted to lower values as the biodegradation time was enhanced, as shown in Figure 3.8 for the nanocomposites with 10% filler content. This effect was especially visible in silica and graphene nanocomposites. The observed reduction in the degradation temperature indicated that the changes happening during the biodegradation process specifically reduced the thermal resistance of chitosan. This also corresponded with the earlier findings of increased crystallinity of PBS as well as preferential degradation of chitosan in the nanocomposites.

Usually, the presence of air in combination with high temperature leads to an earlier breakdown of the chains than in an inert atmosphere. Oxygen promotes chain scissions by the formation of peroxide and hydroperoxide groups, which are highly reactive, thus, favoring the further decomposition reactions [42]. In the TGA-MS

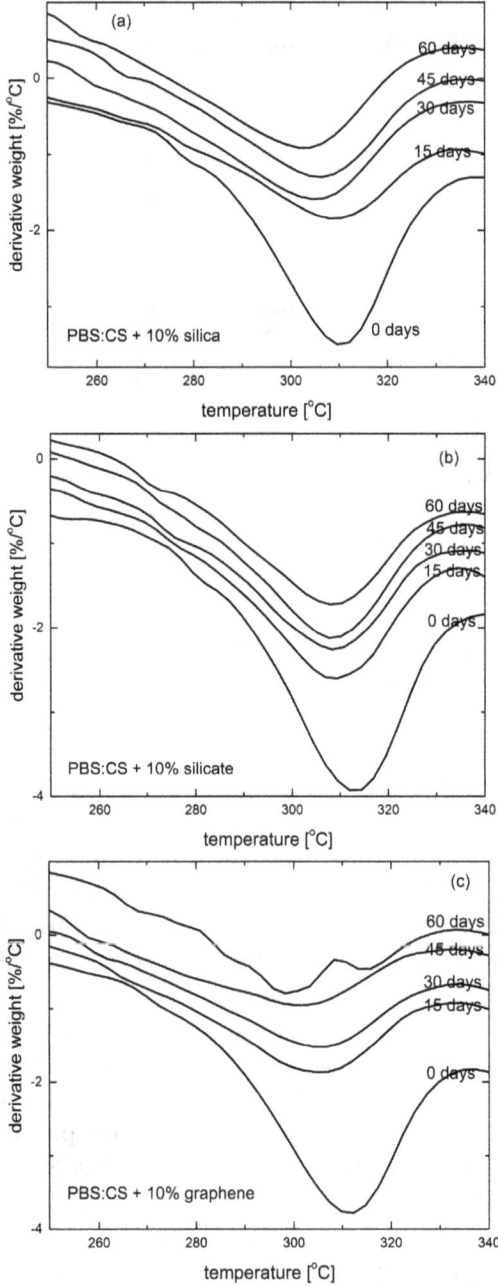

Figure 3.8 TGA thermograms demonstrating the changes in degradation temperature of chitosan as a function of embedding time for (a) silica, (b) silicate and (c) graphene nanocomposites having 10% filler content.

analysis in the air atmosphere for the PBS:CS+5% graphene nano-composite in Figure 3.9, CO_2 (m/z = 44) was observed to evolve around 310 °C, 360-370 °C and 490-510 °C. The peak at 310 °C could be assigned to the degradation of chitosan, whereas the high temperature degradation resulted due to the degradation of graphene. NH_3 (m/z = 17) evolution was also observed during three different stages. The first evolution was linked to the loss of volatile species present in the sample, whereas the second and third evolutions were related to the degradation of chitosan and PBS respectively. Similarly, moisture (m/z = 18) evolution was also observed to evolve during the similar three stages, where the first evolution was due to the loss of free moisture content in the samples. In the nano-composite, the evolution of various components was observed to occur at lower temperature as a function of embedding time during the degradation of both chitosan and PBS. For instance, the samples after 30 d of soil burial exhibited the evolution of these compounds 10 °C earlier than the sample before burial. This is observed because the aerobic oxidation produces functional groups, like –COOH, during the early stages of the biodegradation process that usually have lower thermal stability, thus, resulting in an overall lower degradation temperature. After 30 d of embedding time, the evolution of the components due to degradation was observed to occur at the same temperatures as the samples before burial. It further confirmed that combination of factors like consumption of less thermally stable groups as well as depletion of random amorphous content after 30 d resulted in a shift in the peak degradation temperature back to higher values. It should also be noted that the observed degradation behavior was different from the degradation observed under nitrogen (Figure 3.8), where degradation of the PBS phase was not affected by the embedding time, whereas the chitosan degradation was proportional to the burial time.

Figure 3.10 also compares the TGA-MS spectra of PBS:CS nano-composites with 5% filler content for CO_2 evolution after 30 and 60 d of burial time. The release of CO_2 (and moisture) from the graphene composite was observed to occur at lower temperature, indicating its lower thermal stability as compared to the silica and silicate nanocomposites. Cipiriano et al. [43] reported an increase in the rate of mass loss as the amount of multi-walled carbon nanotubes was enhanced in the composites. This behavior was attributed to an increase in the thermal conductivity of the composite samples. In the present case, this effect may not be responsible for the faster

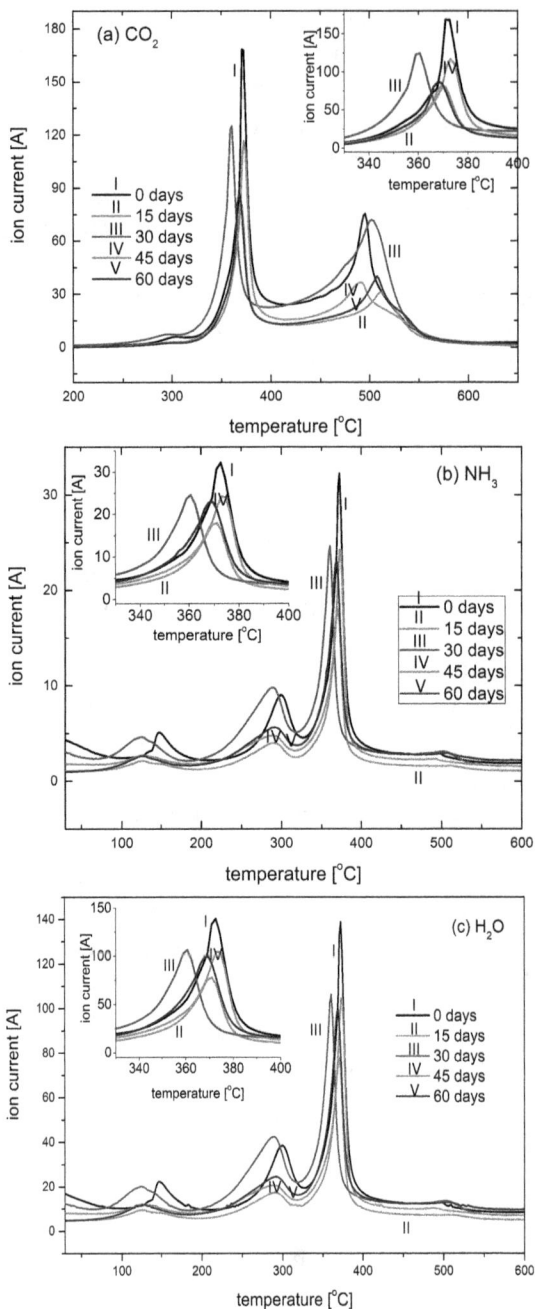

Figure 3.9 TGA-MS spectra of (a) CO_2 (b) NH_3 and (c) H_2O evolution from PBS:CS+5% graphene nanocomposites as a function of soil burial time.

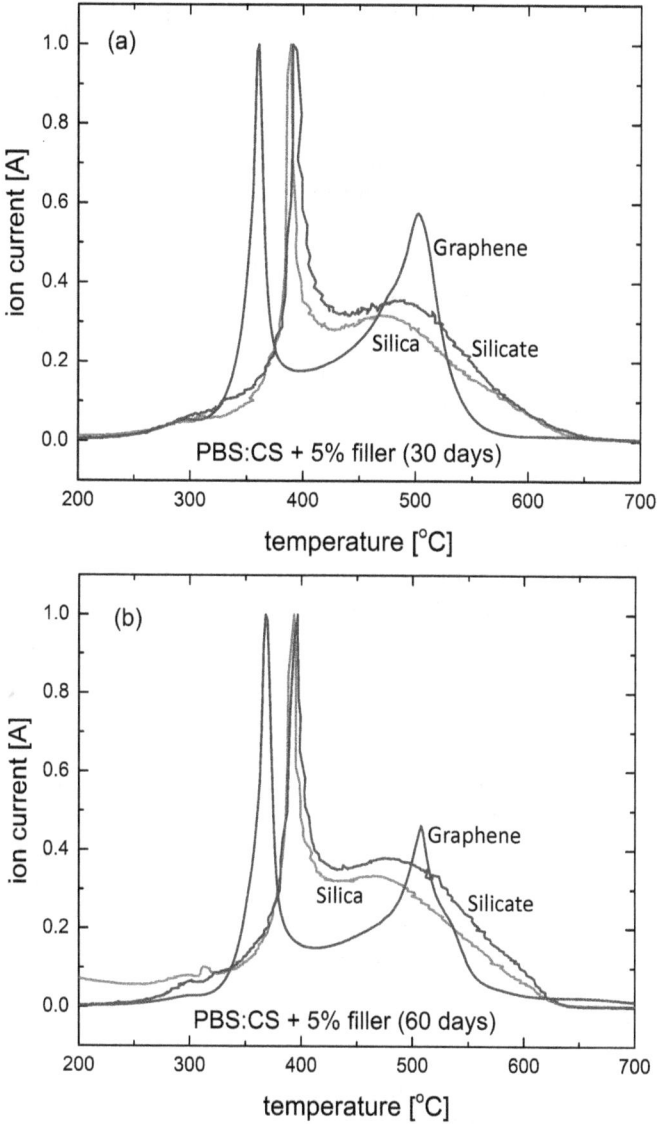

Figure 3.10 TGA-MS spectra for CO_2 evolution from PBS:CS nanocomposites with 5% filler content as a function of soil embedding time; (a) 30 days and (b) 60 days.

thermal degradation observed for graphene based nanocomposites as this effect was not observed under nitrogen atmosphere. Also, the conductivity of the samples did not increase significantly due to graphene addition. Thus, the interaction of functional groups on the graphene surface with air might have resulted in less thermally stable products. The defects/impurities present in the graphene platelets could also contribute to this earlier thermal degradation in the presence of air. The degradation of graphene platelets was also observed by a second broad signal. It should also be noted that though the graphene composites exhibited faster thermal degradation, the biodegradation of these composites was still lower than the corresponding silica and silicate nanocomposites.

For PBS:CS blend, the sample buried for 30 d exhibited reduced peak intensity in the IR spectra indicating occurrence of degradation (not shown). In the nanocomposites, due to comparatively lower extent of biodegradation, the change in IR spectra was not very significant. For silica nanocomposites, the hump around 1090 cm^{-1} corresponding to the Si-O-Si stretching vibration [44] became more prominent in the buried samples. Also, this change in hump was more clearly visible in the samples with higher filler content. In addition, as a function of embedding time, some of the absorption signals corresponding to PBS became broad and diffuse indicating biodegradation. Similar to silica nanocomposites, the silicate nanocomposites did not show any appreciable peak shift or peak broadening. The characteristic depression of the spectrum around 1090 cm^{-1}, which gradually turned into a hump with increasing silicate content confirming the presence of silicate, changed in a similar manner to the silica nanocomposites, but to a lesser extent. Graphene nanocomposites did not show any significant change in peaks. The peak intensity change was observed for the samples with higher graphene content only, thus, indicating that there was reduced extent of biodegradation in the graphene nanocomposites.

3.4 Conclusions

Environmental composting method was used to study the biodegradation properties of PBS:CS nanocomposites with varying amounts of silica, silicate and graphene particles. The composites exhibited lower extent of weight loss with embedding time as compared to PBS:CS blend, however, optical images of the surface of the nanocomposites revealed extensive degradation of the polymer microstructure evident from increased roughness and inter-connected surface cracks. Increasing the amount of silica and silicate particles resulted in increased weight loss due to the loss of polymer strength, however, the graphene nanocomposites exhibited opposite behavior probably due to the specific interaction of graphene plate-lets with the nutrient components. Peak temperature of the melting transition of PBS in the nanocomposites shifted to higher values after soil burial for 30 d which was associated with increased crystallinity due to microbial attack on the amorphous content of the samples. After 30 d, the increase in the melting point stalled which was also confirmed by the corresponding decrease in the crystallinity of the samples. The enhanced crystalline content could also be confirmed through NMR studies where an increase in the intensity of the crystalline bands was observed as a function of embedding time. DSC and AFM analyses also revealed that chitosan was preferentially degraded during the process, which was supported by the decrease in the thermal stability of chitosan in the degraded samples. TGA-MS of the nanocomposites also revealed the degradation of the materials occurring through the evolution of CO_2, H_2O, NH_3 and other components. The evolution occurred at lower temperature for the samples degraded for 30 d. Subsequently, the evolution was observed to occur at higher temperatures due to the depletion of less thermally stable products formed during early stages of biodegradation. Overall, the graphene nanocomposites were observed to have lower thermal stability presumably due to the interaction of surface groups or defects present on surface with air at high temperature. Thus, though the filler hindered the biodegradation of the polymer in the nanocomposites, varying degrees of biodegradation were still achieved depending on the type and amount of filler used. Moreover, the degradation mechanisms were not affected by the type and amount of reinforcement used to generate nanocomposites.

Summary of the Results

Melt processed bio-nanocomposites of poly(butylene succinate) (PBS)-chitosan (CS) generated with varying amounts of silica, alumino-silicate and thermally reduced graphene were analyzed for their biodegradation behavior. The nanocomposite samples were embedded in soil under natural environment for varying periods of time, and the weight loss analysis was complemented with the study of the changes in the surface morphology, crystallinity and thermal stability. Both the type and amount of filler were observed to affect the extent of biodegradation, though no change in biodegradation mechanism occurred. Nanocomposites had, in general, lower extent of weight loss than the pure blend, but the extensive surface roughness and cracks were observed for all systems indicating the initiation of biodegradation. Silica and silicate nanocomposites exhibited higher extent of biodegradation in comparison with the graphene nanocomposites possibly due to the obstructive pathways to microbes in the presence of high aspect ratio graphene platelets. The crystallinity in the pure blend and nanocomposites was observed to increase as a function of embedding time due to the degradation of the random amorphous material in the initial degradation phase. Subsequently, the increase levelled off due to the microbial attack on more organized crystalline content, which was also supported by the reduction in the overall weight loss. An increase in the melting point of PBS with embedding time as well as the depletion of chitosan flakes from the cross-section of the composites in AFM confirmed that chitosan was degraded earlier than PBS. Thermal analysis also indicated faster onset of degradation of chitosan with soil burial time. The degradation studied through TGA-MS also revealed the evolution of H_2O, CO_2 and NH_3, along with other components. The temperature of evolution of these components from the nanocomposites was also affected during different stages of biodegradation.

Acknowledgements

The authors are indebted to Dr. Gisha Luckachan for the solid state NMR analysis.

The definitive version of this work was published earlier in Polymer Composites, 2018, 39(2), 386-397. The work has been reproduced in this chapter with permission from Wiley.

References

1. Lee, S. R., Park, H. M., Lim, H., Kang, T., Li, X., Cho, W. J., and Ha, C. S. (2002) Microstructure, tensile properties, and biodegradability of aliphatic polyester/clay nanocomposites. *Polymer*, **43**, 2495-2500.
2. *Plastic Bags*, Department of the Environment and Energy, Australia (2008). Online: http://www.environment.gov.au/node/21324 [accessed 15th August 2017].
3. *Global Biodegradable Polymers Market to Amount to 2.5 Billion Pounds in 2016*, BCC Research (2011). Online: http://www.bccresearch.com/pressroom/report/code/PLS025D [accessed 19th January 2016].

4. *Biodegradable Polymers: A Review*, Environmental and Plastics Industries Council, Canada (2000). Online: http://www.plastics.ca/ files/file.php?fileid=filehchYqdhbjw&file name=file BIODEGRADEABLE POLYMERS A REVIEW 24 Nov. 20 00. Final.pdf [accessed 10th March 2017].
5. Pagga, U. (1997) Testing biodegradability with standardized methods. *Chemosphere*, **35**, 2953-2972.
6. Lucas, N., Bienaime, C., Belloy, C., Queneudec, M., Silvestre, F., and Nava-Saucedo, J. E. (2008) Polymer biodegradation: Mechanisms and estimation techniques - A review. *Chemosphere*, **73**, 429-442.
7. Mohee, R., Unmar, G. D., Mudhoo, A., and Khadoo, P. (2008) Biodegradability of biodegradable/degradable plastic materials under aerobic and anaerobic conditions. *Waste Management*, **28**, 1624-1629.
8. Yakabe, Y., Nohara, K., Hara, T., and Fujino, Y. (1992) Factors affecting the biodegradability of biodegradable polyester in soil. *Chemosphere*, **25**, 1879-1888.
9. Bordes, P., Pollet, E., and Avérous, L. (2009) Nano-biocomposites: biodegradable polyester/nanoclay systems. *Progress in Polymer Science*, **34**, 125-155.
10. Kumar, A. P., Depan, D., Tomer, N. S. and Singh, R. P. (2009) Nanoscale particles for polymer degradation and stabilization-trends and future perspectives. *Progress in Polymer Science*, **34**, 479-515.
11. Yu, L., Dean, K., and Li, L. (2006) Polymer blends and composites from renewable resources. *Progress in Polymer Science*, **31**, 576-602.
12. Liu, Y. L., Hsu, C. Y., Su, Y. H., and Lai, J. Y. (2005) Chitosan-silica complex membranes from sulfonic acid functionalized silica nanoparticles for pervaporation dehydration of ethanol-water solutions. *Biomacromolecules*, **6**, 368-373.
13. Park, S. B., You, J. O., Park, H. Y., Haam, S. J., and Kim, W. S. (2001) A novel pH-sensitive membrane from chitosan-TEOS IPN; preparation and its drug permeation characteristics. *Biomaterials*, **22**, 323-330.
14. Vijaya, Y., and Krishnaiah, A. (2009) Sorptive response profile of chitosan coated silica in the defluoridation of aqueous solution. *E-Journal of Chemistry*, **6**, 713.
15. Molvinger, K., Quignard, F., Brunel, D., Boissiere, M., and Devoisselle, J. M. (2004) Porous chitosan-silica hybrid microspheres as a potential catalyst. *Chemistry of Materials*, **16**, 3367-3372.
16. Macquarrie, D. J., and Hardy, J. J. E. (2005) Applications of functionalized chitosan in catalysis. *Industrial & Engineering Chemistry Research*, **44**, 8499-8520.
17. Grandfield, K., and Zhitomirsky, I. (2008) Electrophoretic deposition of composite hydroxyapatite-silica-chitosan coatings. *Mate-*

rials Characterization, **59**, 61-67.

18. Ayers, M. R., and Hunt, A. J. (2001) Synthesis and properties of chitosan–silica hybrid aerogels. *Journal of Non-Crystalline Solids*, **285**, 123-127.

19. Correlo, V. M., Boesel, L. F., Bhattacharya, M., Mano, J. F., Neves, N. M., and Reis, R. L. (2005) Properties of melt processed chitosan and aliphatic polyester blends. *Materials Science and Engineering A*, **403**, 57-68.

20. Correlo, V. M., Boesel, L. F., Bhattacharya, M., Mano, J. F., Neves, N. M., and Reis, R. L. (2005) Hydroxyapatite reinforced chitosan and polyester blends for biomedical applications. *Macromolecular Materials and Engineering*, **290**, 1157-1165.

21. Patwary, F., and Mittal, V. (2018) Melt processed PBS/chitosan bio-nanocomposites with silica, alumino-silicate and thermally reduced graphene. In: *Biopolymers Based Advanced Materials*, Mittal, V. (ed.), Central West Publishing, Australia, pp. 1-24.

22. Li, Q., Zhou, J., and Zhang, L. (2009) Structure and properties of the nanocomposite films of chitosan reinforced with cellulose whiskers. *Journal of Polymer Science, Part B: Polymer Physics*, **47**, 1069-1077.

23. Ray, S. S., Okamoto, K., and Okamoto, M. (2003) Structure-property relationship in biodegradable poly (butylene succinate)/layered silicate nanocomposites. *Macromolecules*, **36**, 2355-2367.

24. Rafiee, M. A., Rafiee, J., Wang, Z., Song, H., Yu, Z. Z., and Koratkar, N. (2009) Enhanced mechanical properties of nanocomposites at low graphene content. *ACS Nano*, **3**, 3884-3890.

25. Fan, H., Wang, L., Zhao, K., Li, N., Shi, Z., Ge, Z., and Jin, Z. (2010) Fabrication, mechanical properties, and biocompatibility of graphene-reinforced chitosan composites. *Biomacromolecules*, **11**, 2345-2351.

26. Wang, X., Bai, H., Yao, Z., Liu, A., and Shi, G. (2010) Electrically conductive and mechanically strong biomimetic chitosan/reduced graphene oxide composite films. *Journal of Materials Chemistry*, **20**, 9032-9036.

27. Fang, M., Long, J., Zhao, W., Wang, L., and Chen, G. (2010) pH-responsive chitosan-mediated graphene dispersions. *Langmuir*, **26**, 16771-16774.

28. Rudnik, E. (2008) *Compostable Polymer Materials*, Elsevier Science, USA.

29. Wu, C. S. (2005) A comparison of the structure, thermal properties, and biodegradability of polycaprolactone/chitosan and acrylic acid grafted polycaprolactone/chitosan. *Polymer*, **46**, 147-155.

30. Shogren, R. L., Doane, W. M., Garlotta, D., Lawton, J. W., Willett, J. L. (2003) Biodegradation of starch/polylactic acid/poly (hydroxyester-ether) composite bars in soil. *Polymer Degradation and Stabil-*

ity, **79**, 405-411.

31. Rizzarelli, P., Puglisi, C., and Montaudo, G. (2004) Soil burial and enzymatic degradation in solution of aliphatic co-polyesters. *Polymer Degradation and Stability*, **85**, 855-863.
32. Kim, H. S., Kim, H. J., Lee, J. W., and Choi, I. G. (2006) Biodegradability of bio-flour filled biodegradable poly (butylene succinate) bio-composites in natural and compost soil. *Polymer Degradation and Stability*, **91**, 1117-1127.
33. Jakubowicz, I. (2003) Evaluation of degradability of biodegradable polyethylene (PE). *Polymer Degradation and Stability*, **80**, 39-43.
34. Pandey, J. K., and Singh, R. P. (2001) UV-irradiated biodegradability of ethylene-propylene copolymers, LDPE, and i-PP in composting and culture environments. *Biomacromolecules*, **2**, 880-885.
35. Zhang, Y., Ge, S., Tang, B., Koga, T., Rafailovich, M. H., Sokolov, J. C., Peiffer, D. G., Li, Z., Dias, A. J., McElrath, K. O., Lin, M. Y., Satija, S. K., Urquhart, S. G., Ade, H., and Nguyen, D. (2001) Effect of carbon black and silica fillers in elastomer blends. *Macromolecules*, **34**, 7056-7065.
36. Jang, K. W., Kwon, W. S., and Yim, M. J. (2004) Effects of silica filler and diluent on material properties and reliability of nonconductive pastes (NCPs) for flip-chip applications. *IEEE Transactions on Components and Packaging Technologies*, **27**, 608-615.
37. Kim, H., Abdala, A. A., and Macosko, C. W. (2010) Graphene/polymer nanocomposites. *Macromolecules*, **43**, 6515-6530.
38. Chaudhry, A. U., and Mittal, V. (2013) High-density polyethylene nanocomposites using masterbatches of chlorinated polyethylene/graphene oxide. *Polymer Engineering and Science*, **53**, 78-88.
39. Osman, M. A., Mittal, V., and Suter, U. W. (2007) Poly (propylene)-layered silicate nanocomposites: gas permeation properties and clay exfoliation. *Macromolecular Chemistry and Physics*, **208**, 68-75.
40. McAllister, M. J., Li, J. L., Adamson, D. H., Schniepp, H. C., Abdala, A. A., Liu, J., Herrera-Alonso, M., Milius, D. L., Car, R., Prud'homme, R. K., and Aksay, I. A. (2007) Single sheet functionalized graphene by oxidation and thermal expansion of graphite. *Chemistry of Materials*, **19**, 4396-4404.
41. Sam, S. T., Ismail, H., and Ahmad, Z. (2011) Soil burial of polyethylene-g-(maleic anhydride) compatibilised LLDPE/soya powder blends. *Polymer-Plastics Technology and Engineering*, **50**, 851-861.
42. Corres, M. A., Zubitur, M., Cortazar, M., and Mugica, A. (2011) Thermal and thermo-oxidative degradation of poly(hydroxy ether of bisphenol-A) studied by TGA/FTIR and TGA/MS. *Journal of Analytical and Applied Pyrolysis*, **92**, 407-416.
43. Pandey, J. K., Chu, W. S., Kim, C. S., Lee, C. S., and Ahn, S. H. (2009) Bio-nano reinforcement of environmentally degradable polymer

matrix by cellulose whiskers from grass. *Composites, Part B: Engineering,* **40**, 676-680.

44. Tan, J., Chao, Y. J., Yang, M., Williams, C. T., and Van Zee, J. W. (2008) Degradation characteristics of elastomeric gasket materials in a simulated PEM fuel cell environment. *Journal of Materials Engineering and Performance,* **17**, 785-792.

4

Polyethylene: Degradation and Stabilization

4.1 Introduction

4.1.1 Historical Facts

The production and manufacturing of polyethylene (PE) can be traced back to the early 19th century, as evident from the historical timeline of PE discovery and production shown in Figure 4.1 [1,2].

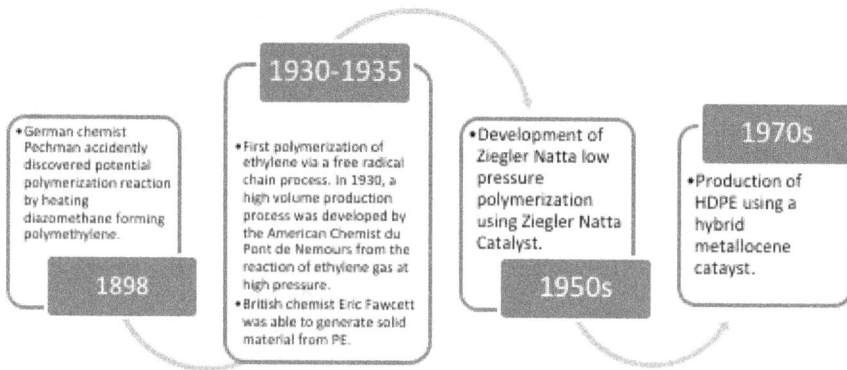

1898
- German chemist Pechman accidently discovered potential polymerization reaction by heating diazomethane forming polymethylene.

1930-1935
- First polymerization of ethylene via a free radical chain process. In 1930, a high volume production process was developed by the American Chemist du Pont de Nemours from the reaction of ethylene gas at high pressure.
- British chemist Eric Fawcett was able to generate solid material from PE.

1950s
- Development of Ziegler Natta low pressure polymerization using Ziegler Natta Catalyst.

1970s
- Production of HDPE using a hybrid metallocene catayst.

Figure 4.1 Historical timeline of polyethylene discovery and production.

4.1.2 Overview of Polyethylene Grades

The synthesis reaction of PE is categorized as addition polymerization which is based on using a radical initiator such as benzoyl peroxide and 2,2'-azo-bis-isobutyrylnitrite (AIBN) to react the ethylene monomer units. Varying operating conditions of PE synthesis reactions can lead to different product characteristics [1,3]. For instance,

Aya Shiraz Shukri AlMasri, Maram Awad, Sawsan Ali and Vikas Mittal, The Petroleum Institute (part of Khalifa University of Science and Technology), Abu Dhabi, UAE*
*Current address: Bletchington, Wellington County, Australia

compressing ethylene gas at pressures as high as 1400-2400 bar at a temperature of 250 °C produces HDPE in the presence of Ziegler-Natta catalyst. High pressure polymerization serves in reducing the side branching. On the other hand, low pressure operation of the polymerization process produces low density PE (LDPE). Co-polymerization with butane and hexane produces medium density PE (MDPE) and linear low density PE (LLDPE), where both poly-mers have shorter side chains and lower extent of branching than LDPE [3,4] (Figure 4.2).

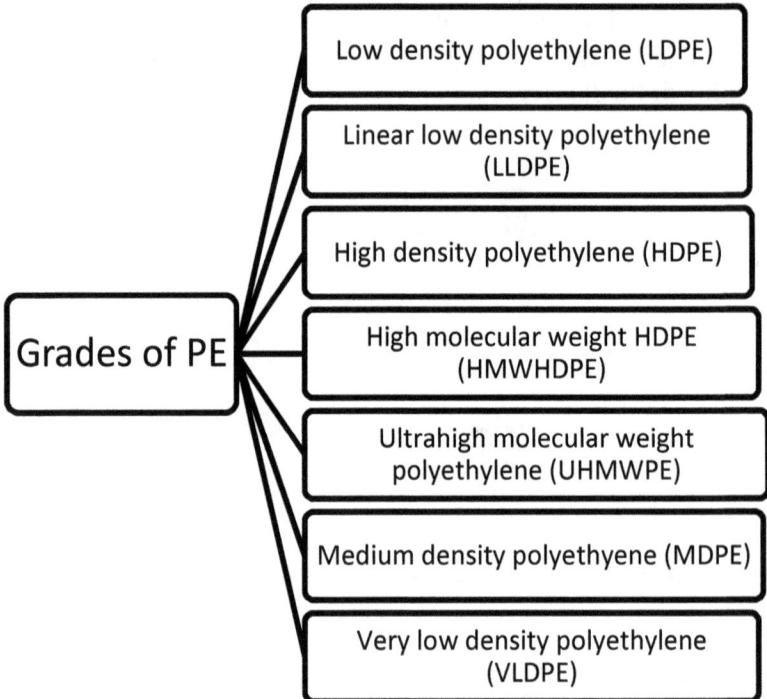

Figure 4.2 Polyethylene grades.

Table 4.1 also provides a brief overview of the key features of various PE grades (LDPE, MDPE, HDPE, LLDPE and UHMWPE). The branching is more prominent in LDPE as compared to HDPE and LLDPE. The higher degree of branching in LDPE reduces its crystal-linity, which also reflects in its physical properties. The increase in the density in other PE grades is mainly attributed to the crystalliz-ing tendency of the polymer chains, which is a strong function of their closely packed chains.

Table 4.1 Brief overview of the different PE grades [2,5,6]

PE grade	Description of chain packing	Rigidity and strength	Crystallinity and density	Flexibility	Major applications
LDPE	Long branched thermoplastic, not closely packed together	Weakest among PE grades	Low density (910-935 kg/m³) and crystallinity	High flexibility	Packaging films for food, plastic bags and containers; electrical insulation; liner material for pools, tanks and ponds
MDPE	Less branching than LDPE, more branching than HDPE	Softer than HDPE and stronger than LDPE	940 kg/m³	Medium flexibility	Gas pipes; carrier bags; screw closures
HDPE	Least branching compared to LDPE and MDPE	Hard, rigid	955-977 kg/m³; more tendency for chain packing; more crystalline than LDPE	Less flexible than LDPE	Squeeze bottles; milk containers; water pipes; garbage containers
LLDPE	Shorter branching than LDPE, this inhibits close packing	Softer and weaker than LDPE	920-930 kg/m³; less tendency to crystallize	Most flexible, more than LDPE	Extruded packaging films; electrical insulations; drain pipes
UHMW PE		Maximum strength and rigidness	NA	NA	Bullet-proof vests; climbing/fishing apparatus; medical

					sectors; joint replacement prostheses in reconstructive surgery

Overall, PE is a semi-crystalline polymer and the degree of crystallinity contributes to increased density which subsequently affects its yield strength, chemical and thermal resistance [3]. The effect of increasing the density of PE on various mechanical and physical aspects is depicted in Table 4.2. It is worth mentioning that the melt viscosity as well as molecular weight (MW) distribution of PE affect the physical and mechanical behavior of the polymer under conditions of contact with chemicals, mechanical energy, tensile loads and high temperature [3,7]. For example, higher melt index caused by a higher PE MW indicates high tensile stress and would require higher energy consumption for heating and softening of the polymer during extrusion process [7].

Table 4.2 Physical and mechanical properties as a function of density, melt index and molecular weight distribution [3,7]

Property	Increase in density will lead to	Increasing melt index will lead to	Broadening of molecular weight distribution will lead to
Tensile strength (at yield)	↑↑	↓↓	
Stiffness	↑↑	↓	↓
Impact strength	↓↓	↓↓	↓↓
Low temperature brittleness	↑↑	↑↑	↓↓
Abrasion resistance	↑↑	↓↓	
Hardness	↑↑	↓	
Softening point	↑↑		
Stress crack resistance	↓↓	↓↓	↑↑
Permeability	↓↓	↑	
Chemical resistance	↑↑	↓↓	

Melt strength		↓↓	↑↑
Gloss	↑↑	↑↑	↓↓
Haze	↓↓	↓↓	
Shrinkage	↓↓	↓↓	↑↑
Melt viscosity	↑↑	↓↓	
Softening point	↑↑	↓↓	↓↓
Surface hardness	↑↑	↓	↓↓
Elongation	↓↓	↓↓	↑↑
Creep resistance	↑↑	↓	↑↑
Flexural stiffness	↑↑	↓	
Flexibility	↓↓		

↑= Increases slightly
↑↑= Increases
↓= Decreases slightly
↓↓= Decreases

4.2 Degradation and Stabilization of PE and Composites

The various degradation mechanisms can be summarized as follows [3]:

- Thermal degradation, which involves chain scission or non-chain scission based processes.
- UV degradation, commonly known as photo-degradation in conjunction with the radiation degradation induced by gamma rays and the presence of impurities which tend to increase radiation absorption.
- Hydrolysis and chemical attack, which causes bond cleavage in the presence of dissolved oxygen in water or by the attack of the chemical agents.
- Machine/mechanical degradation, caused by the imposition of stress and load leading to random chain breakage.
- Microbial biodegradation, which is most prominent in naturally occurring polymers such as cellulose acetate as well as synthetic polyurethane polyesters. A wide variety of biodegradable nanocomposites are PE based, where the applications include packaging foams, disposable food items, food packaging and health care products. Although the biodegradability characteristics of these nanocomposite materials are attractive, however, they lack structural and functional sta-

bility. This interferes with their widespread commercial utilization [8].

- Fire degradation, which can also be manifested by thermal
  degradation, but also requires measuring the heat release
  rate as a function of time using cone calorimetric techniques.

As the stability of polymers is essential for their effective application, several research studies have explored the mechanisms and
kinetic details of degradation and stabilization. The interactions between the various weathering factors such as high temperature,
moisture and UV radiation have been considered for this purpose.
Specifically, the combination of various environmental factors involved in the process of UV and photo-degradation makes the in-
depth understanding of the mechanisms and kinetics very challenging [9].

4.2.1 UV Radiation and Photo-oxidation Degradation

Photo-oxidation is initiated when light is absorbed. UV light is characterized by short wavelength and high photon energy, and it tends
to be absorbed quickly by the polymeric materials. PE, however, is a
non-absorbing polymer, as it does not contain chromophoric species. Without the absorption of light, photochemical reactions cannot proceed [10]. However, oxidized species formed during processing or manufacturing, such as hydroperoxides and peroxides,
create chromophoric defects in PE [10]. The generation of the chromophores tends to accelerate the photo-oxidation reaction by in-
volving specific combinations of the following [3]:

- Reaction of carbonyl group with hydroxyl or hydroperoxides
- Interaction between the peroxy radicals
- Oxidation of hydroxyl group
- Presence of impurities in polyolefins which can be formed
  during synthesis stage such as catalyst metallic residues, aldehydes, ketones, unsaturated vinylene groups (C=C double
  bonds) and acids

Upon exposing PE to UV light, hydroperoxides ROOH or peroxides ROOR autocatalyze the free radical chain oxidation reaction by
accelerating the generation of free radical chain structures. The
weak O-O bond in the hydroperoxides is decomposed to form radi-

cal species which lead to further propagation of photo-oxidation. The products formed are two radicals: a hydroxyl radical and an alkoxy macro-radical, which is an important intermediate species. The intermediate alkoxy macro-radical can react through different pathways [10]:

- Removal of hydrogen without chain cleavage → It forms hydroxyl groups.
- Reaction between the formed radical pair (alkoxy macro-radical and hydroxyl radical) → It forms ketones.
- β-scission with cleavage of the main chain → It forms aldehydes.

The ketones formed in the second possible pathway would then react via photochemical Norrish type I or type II reactions. As these reactions are photochemical in nature, these do not occur under the conditions that are merely thermo-oxidative [10]. Norrish II eventually forms vinyl-type unsaturation radical species which can also possibly react with O_2 to form carbonyl groups within the vinylene structure (ketone species) or can abstract hydrogen to the chain. NII starts with the formation of ketone and vinyl chain species. The ketone will be cleaved into an acetone radical which tends to attack the vinyl chain, transforming it into a free radical vinyl chain. The free radical chain is then oxidized in the presence of O_2, forming a saturated ketone species with carbonyl group attached to the vinyl group, while the vinyl radical chain can be restructured to form a vinylene chain by abstracting a hydrogen atom. On the other hand, Norrish I reactions are observed to form carboxylic acids, esters and lactones.

Stabilization of PE against UV and photo-oxidation can be achieved by one or a combination of the following [11]:

- Carbon black material, which serves as light filtering agent. The efficiency of absorbing the UV light depends on its purity, particle size and degree of dispersion. It may, however, affect the color of the polymer as well as its mechanical properties.
- Hindered amine light stabilizers (HALS), which serve as free radical scavengers and react with radical species during the initiation phase of photo-degradation. The process undergoes the following main steps:

- o Nitroxyl radical is formed from HALS. This radical is stable and captures alkyl radicals before they lead to further degradation.
- o HALS has a regenerative nature which comes from the fact that the formed hydroxyl ethers will tend to combine with the peroxide radicals and regenerate the nitroxyl radical.
- UV absorbers, which convert absorbed UV light into heat. Important examples are triazole, triazine and hydroxyl-benzophenone [10].

The following sections provide a review of the photo-oxidation degradation of some specific PE based materials. Based on the understanding of the degradation mechanisms, stabilization techniques are comprehensively explained.

LDPE Films used as Coverings in Agricultural Services

Photo-oxidation of LDPE is a strong function of the exposure to atmospheric oxygen and can occur at the outer or inner surface depending on the diffusion of oxygen. After a thickness of 20 μm, oxygen diffusion is no longer the rate determining step. Degradation of LDPE can be exacerbated in the presence of oxides like nitrogen dioxide (NO_2) and sulfur dioxide (SO_2) even at temperatures as low as 25 °C. The most commonly known stabilizing additives against such degradation are antioxidants and photo-stabilizers. UV absorbers are commonly used as photo-stabilizers and tend to absorb UV light above a wavelength of 290 nm, preventing it from entering the chromophoric region. The functionality of UV absorbers is more apparent in thick films.

Zinc Oxide (ZnO) and Titanium Dioxide (TiO₂) based HDPE Nanocomposites

Depending on the type of fillers added in the nanocomposites, UV degradation mechanisms widely differ and can subsequently change the role of the filler from counter-degradant to pro-degradant. Additionally, high temperature and O_2 in the degradative environment contribute to the dynamics of the UV degradation process. For example, TiO_2 and ZnO fillers have been observed to serve as photo-catalysts, hence, increasing UV absorption [9].

On the other hand, silica (SiO_2) coated ZnO and TiO_2 nanoparticles serve as counter-degradants and stabilizing agents against UV degradation. HDPE-silica coated ZnO nanocomposite, thus, exhibited increased UV stability. The core mechanism of stabilization is induced by the formation of inert silica shells between ZnO and HDPE which act as a protective physical barrier against photo-oxidation. In fact, the coated nanoparticles of ZnO were found to be more effective in UV stabilization than free radical scavengers such as HALS [9,12].

PE-Multi-walled Carbon Nanotubes (MWCNTs) Nanocomposites

Carbon nanotubes (CNTs) enhance the thermal stability due to the formation of a physical barrier to thermal decomposition. La Mantia and Malatesta [13] and Morlat-Therias *et al.* [14] reported the role of CNTs in providing the UV stability in PE nanocomposites. CNTs absorb the UV photochemical energy and convert it to heat. This acts as a UV filtration mechanism which reduces the transmission of the photochemical energy, but may simultaneously increase the PE bulk temperature and possibly thermo-oxidation. However, optimizing the degradative conditions can make CNTs a UV stabilizing agent rather than a pro-degradant. In most of the tested cases, CNTs were found to have positive effects on the retardation of photo-oxidation [13,14].

HDPE films with 2.5 wt% MWCNTs along with organically modified montmorillonite (OMMT) and SiO_2 nanoparticles were studied by Grigoriadou *et al.* [15]. OMMT and SiO_2 catalyzed photo-oxidation while MWCNTs stabilized photo-oxidation. CNTs can also behave as an antioxidant, increasing the photo-oxidation induction time as well as induction temperature. The surface functionalization of MWCNTs by hydroxyl groups showed an improvement in the stabilization of the HDPE, especially when the MWCNTs content exceeded 1 wt%. The number of walls in MWCNTs also plays a role in the stabilization process [9,16]. Olewnik *et al.* [17] also observed that the decomposition of the organic cations in modified MMT catalyzed the degradation of PE.

HDPE-Calcium Carbonate ($CaCO_3$) Nanocomposites

Calcium carbonate has been observed to be superior to other fillers such as mica and organo-clay in stabilizing the UV degradation and reflecting the ultraviolet light almost completely, thus, protecting the HDPE surface from light induced degradation [9].

HDPE, LDPE, Biodegradable Polyethylene (PE-BIO) and Oxodegradable Polyethylene (PE-OXO) Films under UV-B Radiation

Martinez-Romo *et al.* [18] reported that LDPE can be transformed from its biodegradation resistant form to being biodegradable through photo-oxidation initiated by UV radiation. The general requirement is that the imposed energy should exceed the PE bond energy. The degradation forms reactive vinyl (C=C) and carbonyl (C=O) groups. This also involves depolymerization by chain scission, in absence of quaternary carbon atoms, and crosslinking, in presence of quaternary carbon atoms. Crosslinking occurs as a result of the formed radical species which catalyze the oxidation reaction and eject high energy electrons. The changes in crystallinity can also provide an indication about the comparative susceptibility to degradation by photo-oxidation.

4.2.2 Thermal Degradation

As explained earlier, Norrish reactions are absent in thermo-oxidation degradation processes. The ketones build up as oxidation products while unsaturated vinyl and t-vinylenes groups are not produced. Accordingly, the thermal degradation of PE is mainly driven by chain scission, which is further exacerbated by the combination of O_2 and UV light with high temperature.

The following sections explain different thermal degradation and stabilization aspects experienced during the processing and application stages.

During Processing Stage

Moss *et al.* [19] identified various degradation phenomena for unstabilized Phillips HDPE and unstabilized Ziegler HDPE. Degradation is observed to be either in the form of crosslinking with increase in MW or chain scission. The type of degradation is observed to be related to the nature of catalyst used in the polymerization process. Ziegler catalyst undergoes chain scission, while Phillips catalyst is observed to undergo crosslinking.

Stabilization of HDPE can be achieved by interfering with the degradative components and free radicals as well as by suppressing the autocatalytic reaction. The stabilizing agents can synergistically combine sterically hindered phenols which act as free radical scavengers and phosphite agents which act as hydroperoxide decomposers. For Phillips HDPE, the working principle of these agents is that they reduce the formation of trans-vinylene groups during the multiple extrusion steps. Phosphite hydroperoxide decomposers assist in reducing the amount of phenolic antioxidants required for stabilization. Nitroxyl radical can cap the active free radicals which would reduce the melt flow index due to an increase in MW by crosslinking. For Ziegler HDPE, where chain scission is prominent, the addition of stabilizer leads to a lesser degree of reduction in the vinyl groups. The addition of phenolic free radical scavengers, such as nitroxyl radical inhibitor, helps in preventing the increase in melt flow due to chain scission (reduction in MW) as well as eliminating the formation of vinylene groups [19].

The production of the most PE commodities and packaging items involves thermal processes such as sheet extrusion, film blowing or substrate coating extrusion. These processes utilize high temperatures, thus, degradation during processing is likely to happen. Thermal stresses combined with mechanical/frictional stresses imposed on the polymer melt contribute to degradation as well [20]. Andersson and Wesslen [20] reported that different degradation mechanisms of the PE systems occur at different stages of the extrusion process. Degradation first starts within the barrel, where primary radicals are produced. These radicals auto-oxidize in the presence of oxygen. For HDPE and LLDPE, chain scission is the governing degradation mechanism, while chain scission and crosslinking are prominent for LDPE due to a higher degree of branching and presence of quaternary carbon atoms. When studying the difference in oxidation behavior of HDPE, LDPE and LLDPE without antioxidants, it was concluded that highly branched systems like LDPE crosslink at temperatures under 300 °C and undergo chain scission at higher temperatures [20]. In order to stabilize against the degradation caused by the primary radicals, stabilizing additives such as antioxidants (AO) are used. However, it was found that AO solely are not effective in eliminating the oxidation degradation at the die and specifically in the air gap. Accordingly, the polymer resin should also be nitrogen-saturated, which helps in reducing the oxygen content effectively [20].

During Application Stage

Han *et al.* [21] studied the thermal degradation of polyethylene glycol (PEG) in air and vacuum by placing PEG in an oven at 80 °C for 1000 hours. The authors also studied the stabilizing effectiveness of a phenolic antioxidant 2,2'-methylene-bis(4-methyl-6-*tert*-butylphenol) (MBMTBP). The changes in the melting point with and without MBMTBP were compared for both air and vacuum degradation environments. Upon adding MBMTBP, the melting point was unchanged even after degrading in air. While for vacuum-degraded PEG, no changes in the melting point were noted other than a minor decrease resulting from the antioxidant addition. Accordingly, the functionality of MBMTBP as a thermal degradation stabilizer was observed to increase the thermal stability of PEG without affecting its physical characteristics (i.e. melting point). The mechanism of the stabilization was opined to be:

- Antioxidant MBMTBP captures a PEG radical, or PEG peroxide radical is formed due to the chain breaking cycle.
- MBMTBP forms phenoxy radicals that combines with more MBMTBP.
- PEG thermal degradation terminates because of the formation of the oxidative transformed products of MBMTBP in step 2.

A study by Mohsin *et al.* [4] explored the enhancement of tensile, morphological and thermal stability properties of CNTs/HDPE nanocomposites incorporating MMT. MMT served as a secondary filler and assisted in the adhesion and dispersion of CNTs, along with contributing to the flame retardancy and rheological features [4,22]. The key factor for the thermal resistance enhancement by MMT was reported to be the formation of the interconnected percolative paths which hindered the release of the decomposed volatile compounds.

In another study, Olewnik *et al.* [23] demonstrated the thermal stability of PE nanocomposite with n-heptaquinolinum clay (modified to have a combination of hydrophobic aromatic and alkyl parts as well as hydrophilic parts). It was observed that the modified MMT with n-heptaquinolinum increased the peak melting temperature, enthalpy of melting and crystallinity, thermal degradation temperature, etc.

Khaleel *et al.* [24] developed PE nanocomposites with doped TiO_2/manganese (Mn) nano-powder using *in-situ* polymerization in an inert nitrogen environment. The fillers impeded the alignment of the crystalline structure. The reduction in the degree of crystallinity resulted due to the high branching density induced by the presence of the nanofillers. Accordingly, the melting temperature was also found to be lower. The dispersed doped TiO_2/Mn enhanced the thermal stability, where the TGA (thermogravimetric analysis) curve plateau was broadened, indicating a higher thermal degradation temperature. This was due to an increase in MW resulting from increasing the TiO_2/Mn content. The operating pressure of the synthesis process also affected the thermal stability.

Jeziorska *et al.* [25] studied the thermo-oxidation degradation of the HDPE-SiO_2 nanocomposites. It was observed that the silica particles increased the onset degradation temperature (T_{onset}) by a margin of 8 to 33 °C as compared to pure HDPE. Modified silica, Ag-SiO_2 and Cu-SiO_2, further increased the thermal stability as compared to pure silica. Additionally, silica produced char residue post degradation which acted as a thermal barrier upon exposure to high temperature. One of the important roles of silica in thermally stabilizing HDPE was also reflected in enabling the ablation process, promoted by forming aggregates upon heating. Compatibilizing the HDPE-silica matrix using a compatibilizer enhanced the onset degradation temperature from 392 °C to 420 °C due to the enhanced dispersion of the silica nanoparticles.

The properties of the HDPE-alkylated graphene oxide (MGO) nanocomposites were studied by Gao *et al.* [26] using TGA by employing a heating rate of 10 °C/min. It was found that as the modified graphene oxide content increased to 5 wt%, the thermal stability (T_{onset}) enhanced by 20 °C.

For LDPE films used in agricultural coverings, thermal degradation cannot be de-coupled from the existence of photo-oxidation [3]. These two degradation processes are inter-related and specific measures can be adopted to counter each one of them. Thermal chain scission of LDPE films is inevitable when applied in greenhouse stations. This is linked to the increased content of carbonyl groups promoted by the pesticides used within the greenhouse environment. In order to stabilize the LDPE films used in greenhouse agricultural coverings, commercially manufactured LDPE films are extruded as multilayer films (3 to 7 layers) with 4% to 10% of ethyl vinyl acetate (EVA) [3].

4.2.3 Chemical Degradation

LDPE is resistant to solvolysis and hydrolysis caused by the attack of the active liquids and chemical detergents. However, the combination of high temperature with the solvolysis phenomenon may lead to the initial physical swelling of PE and further dissolution [3].

LDPE is widely used as thin films in agricultural processes as cover for fruit-bearing trees, irrigation pipes, agricultural tunnels and coverings in greenhouses. One degradation mechanism affecting the LDPE agricultural coverings is the attack by sulfur and halogen containing chemicals mainly present in pesticides. The life of the LDPE films used in the agricultural greenhouse coverings depends on the properties of the films, which are governed by the structure and orientation of the chains, use of active agrochemicals or pesticides and their frequency of application, climatic conditions surrounding the greenhouse films such as solar radiation, humidity and debris, ageing of the films, etc.

The effective mitigation measure is to increase the film thickness. Higher thickness tends to contain greater concentrations of stabilizing agents and accordingly contributes to the film durability, along with the UV light and humidity resistance.

4.2.4 Fire Degradation

PE is characterized by a low oxygen index. It is known for its melting and dripping behavior when exposed to high temperatures during fire. Hence, it is essential to enhance the fire resistance, modify its burning behavior to avoid the dripping effects and reduce the heat release rates. Halogenation is one of the useful techniques for enhancing the flame resistance, however, the burning effects of the halogens produce toxic smoke such as furan and brominated dibenzo-p-dioxins (PBDD) [27]. PE nanocomposites with fillers and additives can be considered as a substitute to halogenation. For instance, hydrated alumina liberates water in an endothermic reaction, thereby, cooling the material and increasing the energy needed to maintain the flame. However, it needs high loading and negatively affects the mechanical and physical properties. Phosphorous containing additives either change the degradation mechanism or enhance the char formation which protects the underlying polymer from further degradation. However, these additives are observed to be system-specific.

Liu [28] reported that the presence of magnesium hydroxide (MH) increased the limiting oxygen index (LOI) of LDPE to 25. Furthermore, the replacement of MH with clay was also found to be successful in increasing the LOI to 26. Olewnik *et al.* [17] reported that the presence of OMMT in PE forms an insulating and incombustible char layer post thermal pyrolysis. Regarding thermal stability, T_{onset} was labeled as the point of 10% mass loss (T_{10}). In an air-medium TGA, a higher content of OMMT led to a lower T_{10} and a catalyzed pyrolysis induced by the reaction of the degradation product with clay itself. An interesting study by Sanchez-Olivares *et al.* [27] demonstrated the optimization of the content of the flame retardants embedded in HDPE by the application of ultrasonic treatment during the extrusion and melt intercalation of aluminum trihydroxide (ATH) and zinc borate (ZB) – PE nanocomposites.

4.2.5 Mechanical Degradation

Mechanical degradation of PE is manifested in deteriorating tensile strength, elongation at break and elastic modulus. The elongation at break has been found to be the most sensitive to mechanical degradation and results in an increase in the brittleness of the material. The increase in the carbonyl groups in the material can also lead to enhanced brittleness, thus, aggravating mechanical degradation. The mechanical degradation involves breaking the bonds of the material, thus, releasing the free radicals. Such a non-chemical process also dissipates energy upon rupture, which depends on the duration the material remains strained under the applied mechanical load. Chain scission can occur depending on the inherent strength between the polymer chains. For LDPE, since the linear chains are highly flexible and are not held rigidly against each other, chain slippage will mainly occur upon non-chemical mechanical degradation. However, chain scission will still occur in LDPE and is mainly intensified at the center of the chains rather than at the boundaries.

Crosslinking plays an important role in modifying the mechanical and thermal properties. A study conducted by Alshrah and Janajreh [29] demonstrated the possible measures for recycling the crosslinked PE, which is the base material for making the cable insulation.

Elyashevic *et al.* [30] also investigated the composite systems using porous PE films providing a support matrix with conducting layers of polyaniline (PANI) integrated on the surface and pore walls of PE. PE-PANI composites were developed using *in-situ* polymerization, with PANI loading ranging between 6-8%. Investigations conducted by heating the samples freely without any load showed difference in the thermo-deformational behavior of porous PE versus PE-PANI, with PE-PANI shrinking at higher temperatures. Additional experiments were conducted by heating the samples under loads of 0.02 MPa up to 1.0 MPa and in vacuum. The PE film was observed to shrink and break at 132 °C at any load. The PE-PANI nanocomposite underwent shrinkage till the melting point of PE, while the shrinkage behavior decreased with an increase of load. At a load of 2.0 MPa, the composite did not shrink, rather it underwent elongation until it broke at the melting point.

4.2.6 Water Hydrolytic Degradation

Coloration of PE, increase in brittleness and surface roughness accompanied with erosion are alleviated in the combined presence of moisture and solar radiation. Although PE is a hydrophobic polymer, the presence of water has a degradative effect on LDPE. In an effort to investigate water induced degradation of polymers at room temperature, Massey *et al.* [31] studied the performance of LDPE doped with 4% HALS in comparison with pure LDPE. In X-ray photo-electron spectroscopy (XPS), both samples reported the formation of oxidation products, C-O and C=O groups. The absence of O=C-O groups in the spectra suggested that the hydrolytic degradation mechanism of LDPE did not include chain scission. Upon using HALS with LDPE, the measured oxygen content was found to be less than in the case of unstabilized LDPE. This meant that HALS was effective in reducing the extent of oxidation degradation by capping the radical formed on the –CH bond.

4.2.7 Microbiological Degradation

The applications of synthetic polymers are significantly expanding in the fields of agriculture, transportation and medicine, along with daily life uses. This is due to their lightweight, durability and low cost. In this respect, it is important to develop biodegradable PE to reduce the incineration of waste and production of toxic byproducts. Naturally, PE is non-biodegradable and is resistant to microbial attacks due to its long chains and high MW. Its hydrophobic nature and 3-D structure do not allow colonization of microorganisms. Therefore, PE should be decomposed into smaller fragments to be accessible for microorganisms as a food source. In a recent study, Mehmood *et al.* [32] reported the biodegradation of LDPE modified with TiO$_2$.

4.3 Conclusions

The chapter presents a detailed explanation of various PE degradation phenomena. The essential focus of the study is also to assess the corresponding stabilization needed to counter the various degradation processes. Both neat PE as well as PE nanocomposite materials have been studied for this purpose.

References

1. Batra, K. (2013) *Role of Additives in Linear Low Density Polyethylene Films,* Indian Institute of Technology Kharagpur, India. Online: https://www.slideshare.net/kamalbatra111/polyethylene-pe [accessed 10th May 2019].
2. Khanam, P. N., and Al Maadeed, M. A. (2015) Processing and characterization of polyethylene-based composites. *Advanced Manufacturing: Polymer & Composites Science,* **1**, 63-79.
3. Dilara, P. A., and Briassoulis, D. (2000) Degradation and stabilization of low-density polyethylene films used as greenhouse covering materials. *Journal of Agricultural Engineering Research,* **76** (4), 309-321.
4. Ali Mohsin, M. E. Arsad, A., and Alothman, O. Y. (2014) Enhanced thermal, mechanical and morphological properties of CNT/HDPE nanocomposite using MMT as secondary filler. *International Journal of Chemical and Molecular Engineering,* **8** (2), 117-120.
5. Park, S., and Choi, I. S. (2009) Production of ultrahigh-molecular weight polyethylene/pristine MWCNT composites by half-titanocene catalysts. *Advanced Materials,* **21**, 902-905.
6. HDPE, LDPE, LLDPE - What's the Difference Anyway? Global Plastic Sheeting, USA (2018). Online: https://www.globalplasticsheeting.com/hdpe-vs-lldpe-vs-ldpe; [accessed 15th November 2018].
7. *History and Physical Chemistry of HDPE.* Online: Gabrielle, H. History and Physical chemistry of HDPE [accessed 9th May 2019].
8. Schmidt, D., Shah, D., and Giannelis, E. P. (2002) New advances in polymer/layered silicate nanocomposites. *Current Opinion in Solid State and Materials Science,* **6**(3), 205-212.
9. Nechifor, M. (2016) Factors influencing the photochemical behavi-

or of multicomponent polymeric materials. In: *Photochemical Behavior of Multicomponent Polymeric-based Materials*, Rosu, D., and Visakh, P. M. (eds.), Springer, Germany, pp. 21-65.

10. Gardette, M., Thérias, S., Gardette, J., Janecska, T., Földes, E., and Pukánszky, B. (2013) Photo- and thermal oxidation of polyethylene: comparison of mechanisms and influence of unsaturation content. *Polymer Degradation and Stability*, **98**(11), 2383-2390.

11. Kasza, G. (2013) *Thermal, Antioxidative and Photochemical Stabilization of Polymers: Low Molecular Weight versus Macromolecualr Stabilizers*, Polymer Institute of the Slovak Academy of Science, Slovakia. Online: https://slideplayer.com/slide/4610618/ [accessed 19th April, 2019].

12. Ammala, A., Hill, A. J., Meakin, P., Pas, S. J., and Turney, T. W. (2002) Degradation studies of polyolefins incorporating transparent nanoparticulate zinc oxide UV stabilizers. *Journal of Nanoparticle Research*, **4**(1-2), 167-174.

13. La Mantia, F. P., and Malatesta, V. (2009) Photo-oxidation behavior of polyethylene/multi-wall carbon nanotube composite films. *Polymer Degradation and Stability*, **94**, 162-170.

14. Morlat-Therias, S., Fanton, E., Gardette, J., and Dubois, P. (2007) Polymer/carbon nanotube nanocomposites: Influence of carbon nanotubes on EVA photo degradation. *Polymer Degradation and Stability*, **92**, 1873-1882.

15. Grigoriadou, I., Chrissafis, K., and Bikiaris, T. G. (2011) Effect of different nanoparticles on HDPE UV stability. *Polymer Degradation and Stability*, **96**, 151-163.

16. Lonkar, S. P., Kushwaha, O., and Singh, G. (2012) Self photostabilizing UV-durable MWCNT/polymer nanocomposite. *RSC Advances,* **2**, 12255-12262.

17. Olewnik, E., Garman, K., and Czerwinski, W. (2010) Thermal properties of new composites based on nanoclay, polyethylene and polypropylene. *Journal of Thermal Analysis and Calorimetry*, **101**, 323-329.

18. Martínez-Romo, A., González-Mota, R., Soto-Bernal J. J., and Rosales-Candelas, I. (2015) Investigating the degradability of HDPE, LDPE, PE-BIO, and PE-OXO films under UV-B radiation. *Journal of Spectroscopy*, **2015**, doi: 10.1155/2015/586514.

19. Moss, S., and Zweifel, H. (1989) Degradation and stabilization of high density polyethylene during multiple extrusions. *Polymer Degradation and Stability*, **25**, 217-245.

20. Andersson, T., and Wesslén, B. (2003) Degradation of LDPE, LLDPE and HDPE in Film Extrusion. *TAPPI European PLACE Conference,* **2**, 333-359.

21. Han, S., Kim C., and Kwon, D. (1997) Thermal/oxidative degradation and stabilization of polyethylene glycol. *Polymer*, **38**(2), 317-

323.

22. Ali Mohsin, M. E., Arsad, A., Fouad, H., Jawaid, M., and Alothman, O. Y. (2014) Enhanced mechanical and thermal properties of CNT/HDPE nanocomposite using MMT as secondary filler. *AIP Conference Proceedings*, **1599**, 206-209.

23. Olewnik, E., Garman, K., Piechota G., and Czerwinski, W. (2012) Thermal properties of nanocomposites based on polyethylene and n-heptaquinolinum modified montmorillonite. *Journal of Thermal Analysis and Calorimetry*, **110**, 479-484.

24. Abdul Khaleel, S. H., Bahuleyan, B. K., Masihullah J., and Al-Harthi, M. (2011) Thermal and mechanical properties of polyethye-lene/doped-tio$_2$ nanocomposites synthesized using in situ polymerixzation. *Journal of Nanomaterials*, **2011**, doi: 10.1155/2011/964353.

25. Jeziorska, R., Szadkowska, A., Zielecka, M., Wenda, M., and Kepska, B. (2017) Morphology and thermal properties of HDPE nanocom-posites: Effect of spherical silica surface modification and compat-ibilizer. *Polymer Degradation and Stability*, **145**, 70-78.

26. Gao, W., Lu, Y., Chao, Y., Ma, Y., Zhu, B., Jia, J., Huang, A., Xie, K., Li, J., and Bai Y. (2017) Performance evolution of alkylation graphene oxide reinforcing high-density polyethylene. *The Journal of Physical Chemistry*, **121**, 21685-21694.

27. Sanchez-Olivares, G., Sanchez-Solis, A., Calderas, F., Medina-Torres, L., Herrera-Valencia, E., Castro-Aranda, J., Manero, O., Blasio, A., and Alongi, J. (2013) Flame retardant high density polyethyelene optimized by on-line ultrasonic extrusion, *Polymer Degradation and Stability*, **98**, 2153-2160.

28. Liu, S. (2014) Flame retardant and mechanical properties of poly-ethylene/ magnesium hydroxide/montmorillonite. *Journal of In-dustrial and Engineering Chemistry*, **20**, 2401-2408.

29. Alshrah, M., and Janajreh, I. (2013) Mechanical Recycling of Cross-link Polyethylene: Assessment of Static and Viscoelastic Proper-ties. *Renewable and Sustainable Energy Conference (IRSEC)*, doi: 10.1109/IRSEC.2013.6529674.

30. Elyashevich, G. K., Sidrovich, A. V., Smirnov, M. A., Kuryndin, I. S., Bobrova, N. V., Trchova, M., and Stejskal, J. (2006) Thermal and structural stability of composite systems based on polyaniline de-posited on porous polyethylene films. *Polymer Degradation and Stability*, **91**, 2786-2792.

31. Massey, S., Adnot, A., Rjeb, A., and Roy, D. (2007) Action of water in the degradation of low-density polyethylene studied by X-ray pho-toelectron spectroscopy. *eXPRESS Polymer Letters*, **1**(8), 506-511.

32. Mehmood, C. T., Qazi, I. A., Hashmi, I., Bhargava, S., and Deepa, S. (2016) Biodegradation of low density polyethylene (LDPE) modi-fied with due sensitized titania and starch blend using *Stentroph-*

omonas pavanii. International Biodeterioration & Biodegradation, **113**, 276-286.

5

Epoxy: Understanding Degradation and Stabilization

5.1 Introduction

Epoxy resins are either low molecular weight pre-polymers or cured polymeric networks. Historically, epoxy compounds have been discovered by Prileschaiev in 1909, and the commercial products including adhesives and casting resins were marketed in USA in 1946 [1,2]. Epoxy resins exhibit outstanding features such as high thermal and chemical stability as well as adhesion and flame resistance depending on the processing method. Nowadays, epoxies are one of the most widely used thermosetting resins in the fields of adhesives, coatings, semiconductor encapsulation, hardware components, electronic circuit board materials and aerospace.

Polymer degradation is a critical issue as it causes the discoloration, embrittlement and material deterioration [3]. Thermal degradation, oxidation and UV irradiation are the major routes for the degradation processes. The degradation of epoxy takes place largely under high temperature and oxidative conditions. Exposure of UV light combined with oxygen leads to a synergetic effect. Stabilization process depends on the type of stabilizers that act as hydroperoxide decomposers, radical scavengers, UV absorbers and reaction modifiers such as fire retardants. The stabilizer usually reacts in amorphous solid phase or highly viscous state [4-6]. The mobility of the stabilizer is required to be enough to reach the reaction site, but not too high to be lost through the surface. The molecular weight of the commercial stabilizers is usually adjusted in the range of 200-2000 g/mol.

In this chapter, the degradation behavior of epoxy is briefly studied by focusing on the mechanism of induced damage. Stabilization of epoxy with additives to overcome the degradation phenomena is also briefly introduced together. Such insights into the degradation and stabilization phenomena are required for expanding the commercial uses of epoxy.

Hyemin Lee, Zhou He, Yuting Li and Vikas Mittal*, The Petroleum Institute (part of Khalifa University of Science and Technology), Abu Dhabi, UAE
*Current address: Bletchington, Wellington County, Australia

5.2 Understanding the Degradation and Stabilization of Epoxy

5.2.1 Thermal Degradation

Two major factors have an effect on the thermal degradation of epoxy: the structure of epoxy and service conditions. Monomer structure and curing agent used to generate the network determine the thermal behavior. Polymers with aromatic rings in the structure are generally considered to be thermally stable. Similarly, the aromatic epoxy resin is much more stable than the aliphatic one because of the ring structure. Curing agent is generally adopted by considering application conditions, workability and other factors. Christiansen *et al.* [7] revealed that the epoxy resins cured with novolacs had excellent thermal stability, while the epoxy resins cured with dicyandiamide (DICY) performed poorly. Regarding service conditions, epoxy has overall high thermal stability. Degradation generally initiates above 150 °C. The investigation of the thermal ageing stability of epoxy has been mostly performed at high temperatures accordingly. It has been reported that the unfilled epoxy shows short-term thermal resistance at 220-250 °C and rapid degradation over 250 °C. The duration of the exposure of heat and partial oxygen pressure also contributes to the thermal degradation. Also, it is important to notice that the thermal degradation of epoxy resins at relatively low temperature is controlled by diffusion rather than reaction.

Thermal Degradation Mechanism of Epoxy

Degradation can proceed by chain scission and crosslinking, depending on the characteristics of epoxy [8]. It has been understood that chain scission usually contributes to degradation, whereas crosslinking can enhance stability. One of the indicators to assess the degradation mechanism is the glass transition temperature (T_g). The extent of chain scission can be deduced by measuring the decrease of T_g. Reversely, the crosslinking development in the molecule is assumed when T_g increases upon thermal ageing.

   Thermal degradation of the epoxy resins at high temperatures is generally initiated by the formation of vinylene ethers on the backbone when secondary alcohol is dehydrated [9]. The dehydration causes the allylic ether bond to be weakened, which leads to chain scission in the structure. Depending on the curing agent, the stability of the bond varies. The epoxy resins cured by anhydrides are observed to have better thermal stability than the epoxy resins cured by amines. It is assumed that the nucleophilic nature of the amine groups within the network leads to a weak allylic amine C-N bond as compared to the C-O bond. Amine tends to undergo either volatilization or charring as solid residue as the following step. Secondary chain scission of the amine group (C-NH) has been rarely observed in the amine-cured epoxy resins [10].

The study by Zahra *et al.* [11] introduced the possible chain scission route for diglycidyl ether of bisphenol A (DGEBA) cured by jeffamine (POPA) or polyamidoamine (PAA) in low temperature range from 70 to 150 °C by analyzing the IR spectra. It was suggested that the amide group was formed as a major product of oxidation induced by the attack of α amino carbon group of DGEBA, regardless of the curing agents. In the long term experiment for almost a month at 150 °C, the T_g of DGEBA cured by POPA kept decreasing, while the T_g of DGEBA cured by PAA showed increasing tendency, which could be interpreted as predominant chain scission with DGEBA/POPA system and predominant crosslinking reaction with DGEBA/PAA system with time.

Ernault *et al.* [12] tested the steric effect of diamine hardener on the thermal oxidation mechanism of the DGEBA based epoxy resin. The oxidation mechanism of DGEBA resins cured with cycloaliphatic (isophorone diamine; IPDA) or linear aliphatic (trioxa-tridecanediamine or TTDA) diamine hardener was deduced by tracking the T_g variations. It was observed that DGEBA/IPDA resin mainly took the chain scission route with faster oxidation rate as compared to EGEBA/TTDA system that presumably underwent crosslinking in some parts.

In the case of epoxy-clay nanocomposites, it is suggested that the thermal degradation behavior of the DGEBA-montmorillonite (MMT) composites relies on mainly three factors, including the clay loading, structure of nanocomposite and purge gas [13]. Pandey *et al.* [14], however, argued that MMT could influence in both positive and negative ways as the reactivity of the epoxy resin with the curing agent decreased when MMT was added, whereas MMT conferred insulating function to impart protection from oxygen [14]. For example, the exfoliated nanocomposites exhibit better barrier properties and thermal stability than the corresponding intercalated ones as the interference effect is dominant for the intercalated nanocomposites, while the insulation effect is observed to be largely dominant for the exfoliated ones [15].

Thermo-oxidative Degradation Behavior of Epoxy

Degradation behavior of epoxy under thermo-oxidative conditions has been widely studied. Intrinsic characteristics of epoxy networks including the monomer structure and curing agent play an important role in the thermo-oxidation of epoxy. It is also important to notice that the degradation is highly diffusion limited [16]. Discoloration and the formation of oxidized layer with weight loss and density decrease on the surface of epoxy appear as a result, thus, leading the surface layer to shrink and possibly crack [17].

Other reports have pointed out that the oxidative reaction in epoxy resins is limited to the surface as the oxidized layer provides protection from further oxidation. One recent study concluded that the embrittlement in the amine-cured epoxy resin was not directly related to T_g or weight loss, but was governed by the oxidation of hydroxypropylether groups in the DGEBA segment, regardless of the curing agent [18]. As per the study, the gap between T_g and imposed temperature did not show relevance with embrittlement, but the network undergoing crosslinking exhibited significant mass loss [18].

Several recent studies have reported the superficial oxidized layer development on epoxy materials [19-21]. For instance, Yang *et al.* [21] studied the oxidation of the anhydride-cured epoxy resin under thermal conditions (130-160 °C) for 30 days and observed noticeable free volume fraction decrease in the skin layer relative to the bulk sample. The oxidized skin thickness was observed to be less than 100 μm. The comparison of the compressive behavior of the epoxy resin and 3D carbon fibers/epoxy braided composite was also performed after thermal aging by Zhang *et al.* [20]. Stress-strain curves of the resin and 3D composite indicated that the curve of the epoxy resin became stable after 1 day, however, the composite showed continuous decrease with time. It could be assumed that the epoxy resin was protected by the oxidized layer on the surface, while the interface cracking in the composite led to crack propagation and oxygen penetration into the core.

Thermo-oxidative reaction of the epoxy resins under gamma irradiation has received attention because of the application in nuclear industry as coatings. Most of recent studies have focused on the oxidation reaction of epoxy with relatively low glass transition temperatures (typically 50 °C $\leq T_g \leq$ 80 °C). Galant *et al.* [22] studied the thermal and radio-oxidation of the epoxy coating with low T_g epoxy-amine networks under low temperature conditions (50-110 °C) at

200 Gy/h and observed that the tertiary amine group was vulnerable to oxidation regardless of the type of exposure (thermal or radio), thus, resulting in the amide group formation. Variation of the gamma ray dosage accelerated the oxidation rate at low level (50 Gy/h), but the rate became constant at high dosage (200 and 2000 Gy/h). Comparison of the DGEBA/PAA and DGEBA/POPA systems at low level of irradiation indicated that the DGEBA/POPA system developed cross-linking, while the DGEBA/PAA system went through chain scission reaction [23]. Another degradation study of the epoxy-amine network showed carbonyl and hydroperoxide formation under oxygen and gamma irradiation conditions [24]. The study also reported that chain scission was predominant in oxidative degradation.

5.2.2 Fire Retardation

Highly flammable nature of the epoxy based materials has been an issue for extended applications in many fields such as electronics, automotives, aircrafts, etc. The decomposition of the epoxy resins is affected by the reactive curing agents and fire retardant additives. Reactive curing agents such as amines, anhydrides and phenolic groups have a significant impact on flammability. Generally, the amine-cured epoxy resins show better flame stability as compared to the anhydride-cured epoxy resins. The ratio of carbon to oxygen in the structure is closely linked to the combustion behavior. Regarding the inorganic fillers, such as alumina or silica, Paterson-Jones *et al.* [25] also suggested that the catalytic reaction by the fillers in the epoxy resin could accelerate thermal decomposition.

The decomposition mechanism of brominated epoxy resin (DGEBA/DDS) was suggested by Luda *et al.* [26]. It was observed that the brominated part decomposed first, followed by the decomposition of the other parts and char formation. Halogenated compounds are, however, substituted by alternatives because of the toxic gas generation issue. The phosphorous compounds have received attention to replace the halogenated compounds as fire retardants. It is suggested that the flame retardant characteristic is provided by the phosphorous compounds through modification of the thermal decomposition during ignition, which yields carbonaceous char rather than CO and CO_2. The char acts as a physical barrier for combustible gases as well as thermal insulation in the decomposition process. The study by Liu *et al.* [27] proposed that the thermal degradation mechanism of phosphorous containing epoxy (bis-(3-glycidyloxy) phenylphosphine oxide (BGPPO) and 4,4'-diaminodiphenylsulfone (DDS) initiated from phosphorous-phenol bond. Dehydration by P-O-C chain scission and elimination of propyl groups took place during degradation, subsequently leading to char yields with high phosphorous contents.

5.2.3 UV Degradation

Epoxy coatings are widely used as long-term protective layers to safeguard the metal surfaces due to high mechanical properties, adhesion and solvent resistance. The coatings are often exposed to oxidative and UV irradiation conditions, which can lead to the degradation of the material. Many research studies have been carried out to understand the degradation mechanism under UV, such as pyrolysis or photo-oxidation, and to establish the pathways to improve the light resistance of epoxy.

The photo-oxidation mechanism of epoxy maleate of bisphenol A (EMBA) through hydroperoxide intermediates has been suggested by Rosu *et al.* [28]. The degradation was observed to be initiated by the free radical reaction and excitation of double bonds in EMBA at initial stage. Some crosslinked structure formation in EMBA was also suspected at an early stage of photolysis. Hydrogen peroxide formation as secondary reaction led to chain scission in the structure, producing etheric aromatic group.

In contrast, the properties of epoxy thermosets can be improved in many cases at initial stages of UV exposure owing to the residual crosslinking before degradation [29]. Liu *et al.* [30] studied the microstructure evolution of the epoxy coatings, polyamide-cured DGEBA, under UV irritation (55 mW/cm²) for 399 h and revealed that the epoxy coatings could be reinforced under UV light. Upon exposure, the development of a highly crosslinked structure was observed on the surface of the epoxy coatings as a result of recombination, which enhanced the water resistance of the coatings.

5.2.4 Epoxy Stabilization

Epoxy has a wide range of applications. The stability and non-flammability of epoxy at the processing temperature and outdoor condition are essential properties. The purpose of stabilization is to conserve the original properties during various service conditions. Stabilizers are, thus, added to slow down the deterioration, discoloration and embrittlement of epoxy [3]. Stabilizers are mostly dispersed in the amorphous region, and the morphology affects the stabilization process. The stabilizer optimization is of immense importance as excess mobility can lead to the loss of the material. Frequently used stabilizer types for epoxy are antioxidants, light stabilizers, fire retardants, etc.

Antioxidants

Oxidation is one of the major processes leading to polymer degradation. Antioxidants function by combining with free radicals to prevent the progress of degradation. Low molecular weight compounds are usually employed as free radical scavengers. However, majority of the research studies on epoxy materials have focused on the thermal decomposition and development of flame retardants, rather than antioxidants. Recently, eugenol-based epoxy materials have been designed as effective antioxidant [31]. Burton [32] also studied the protective effect of the commercial antioxidants such as tris(nonylpheny1) phosphite on triethylenetetraamin-cured DGEBA at 125 °C. No noticeable protection was observed, thus, explaining that the predominant degradation mechanism of the aliphatic amine-cured epoxy resins was not controlled by free radical processes.

Ghasemi-Kahrizsangi *et al.* [33] investigated the corrosion behavior of the modified nano-carbon black (CB)/epoxy coatings at 65 °C. Addition of CB in the coatings resulted in better corrosion resistance owing to the improved barrier against ion diffusion. Wei *et al.* [34] also applied carbon black to obtain protection for mild steel in 3% NaCl. CB-filled fusion bonded epoxy coatings were observed to affect the corrosion behavior in a positive way (with CB weight fraction above percolation threshold).

Light Stabilizers

Light stabilizers include UV absorbers, quenchers of excited states, hydroperoxide decomposers and free radical scavengers. Many materials are observed to have several of such functions, e.g. carbon black. Light stabilizers interfere with the light-induced reactions and achieve inhibition. Dispersion, particle size and type of light stabilizers are important factors affecting their function [35].

Carbon black is known for its effective UV stabilizing efficiency as well as antioxidant effect in the polymer industry [36,37]. CB/epoxy nanocomposite coatings were investigated by Ghasemi-Kahrizsangi *et al.* [38] under UV irradiation to investigate the effect of CB nano-particles. CB interfered with the formation of micro-cracks and penetration of ionic species by absorbing UV. The epoxy coatings with 2.5 wt% CB exhibited effective degradation resistance after 2000 h exposure and 24 h immersion in 3.5 wt% NaCl.

The addition of clay in epoxy is often observed to improve thermal, viscoelastic and mechanical properties of the polymer, along with a positive influence on the degradation mechanism [39]. However, the presence of transition metals in MMT and their interaction with UV light have also been observed to accelerate the photo-oxidation process [40]. Tcherbi-Narteh [41] also reported that the degradation mechanism varies depending on the clay structure and loading.

Fire Retardants

Flame retardants modify the pyrolysis or oxidation of polymers by slowing or inhibiting the reactions. Development of fire retardants for epoxy has seen accelerating trend in order to extend its applications to various industries, especially electronics and aerospace. There are mainly two ways to impart the fire retardant properties to the epoxy resins. One is the incorporation of the fire retardant additives and the other one involves copolymerization with reactive fire retardants [10]. Halogenated compounds, either as reactive co-reactants or additives, have been widely used as flame retardants traditionally. The release of toxic and corrosive gases during combustion, however, was the major drawback imposing fire hazards and environmental concerns [42]. Thus, the attention nowadays has been focused on the development of halogen-free flame retardants.

Phosphorous derivatives are promising alternative fire retardants for cured epoxy resins, especially 9,10-dihydro-9-oxa-10-phospha-phenanthrene-10-oxide (DOPO) and its derivatives. The efficiency of the phosphorous derivatives originates from their tendency to react with the hydroxyl groups present abundantly in the epoxy structure [10]. The phosphorous structure can also influence the flame retardant performance by affecting the char layer of epoxy [43]. DOPO is mostly used because of its high thermal stability, reactivity and resistance to oxidation [44,45]. Phosphorous derivatives can be applied in epoxy as both additives and reactive co-monomers. The enhanced phosphorous content in the epoxy resin was also observed by Jain *et al.* [46] to improve the limiting oxygen index (LOI). It has been pointed out, however, that organo-phosphorous alone is not capable of conferring the flame retardation properties. The epoxy systems with multiple flame retardants have received attention accordingly.

Yang *et al.* [47] synthesized the flame retardant with phosphorus/nitrogen/boron components and blended it with DGEBA. The results indicated 45.1-72.8% improvement in char yield. Yang *et al.* [48] also conducted copolymerization of DGEBA and phosphorous/nitrogen containing flame retardant (DOPO and N-(4-hydroxyphenyl) maleimide (HPM)). The authors observed improved LOI and decreased total heat release [48]. Boron and silicon containing epoxy also showed a significant reduction in peak heat release rate and total heat release by 69% and 46% respectively as compared to the epoxy resin without flame retardants [49]. With 1.5 wt% boron and 0.5 wt% silicon, the LOI was observed to be 30.5%.

5.3 Conclusions

The thermal decomposition at high temperature with oxidation is observed to be the major pathway for the epoxy resin degradation. The degradation process is generally controlled by oxygen diffusion rather than reaction. Regarding fire retardation, the ratio of carbon to oxygen in the structure is closely linked to the combustion behavior. LOI and char yield are useful indicators to analyze the effect of fire retardants. For stabilization, carbon black has been widely applied for epoxy due to its multitude of properties.

References

1. *Thermal Degradation of Polymer Blends, Composites and Nanocomposites*, Visakh, P. M., and Yoshihiko, A. (eds.), Springer International Publishing, Switzerland (2015).
2. Hamerton, I. (1996) *Recent Developments in Epoxy Resins*, Smithers Rapra, USA.
3. Allen, N. S., and Edge, M. (1992) *Fundamentals of Polymer Degradation and Stabilization*, Elsevier, USA.
4. Minagawa, M. (1989) New developments in polymer stabilization. *Polymer Degradation and Stability*, **25**, 121-141.
5. Saba, N., Tahir, P., and Jawaid, M. (2014) A Review on potentiality of nano filler/natural fiber filled polymer hybrid composites. *Polymers*, **6**(8), 2247-2273.
6. Liu, S., Chevali, V. S., Xu, Z., Hui, D., and Wang, H. (2017) A review of

extending performance of epoxy resins using carbon nanomaterials. *Composites, Part B: Engineering,* **136**, 197-214.

7. Christiansen, W., Shirrell, D., Aguirre, B., and Wilkins, J. (2001) Thermal Stability of Electrical Grade Laminates based on Epoxy Resins. Proceedings of *IPC Printed Circuits EXPO*, USA.

8. Mohan, P. (2013) A critical review: The modification, properties, and applications of epoxy resins. *Polymer-Plastics Technology and Engineering,* **52**, 107-125.

9. Levchik, S. V., Camino, G., Luda, M. P., Costa, L., Costes, B., Henry, Y., Morel, E., and Muller, G. (1995) Mechanistic study of thermal behavior and combustion performance of epoxy resins: I homopolymerized TGDDM. *Polymers for Advanced Technologies,* **6**, 53-62.

10. Levchik, S. V., and Weil, E. D. (2004) Thermal decomposition, combustion and flame-retardancy of epoxy resins - A review of the recent literature. *Polymer International,* **53**, 1901-1929.

11. Zahra, Y., Djouani, F., Fayolle, B., Kuntz, M., and Verdu, J. (2014) Thermo-oxidative aging of epoxy coating systems. *Progress in Organic Coatings,* **77**, 380-387.

12. Ernault, E., Richaud, E., and Fayolle, B. (2016) Thermal oxidation of epoxies: Influence of diamine hardener. *Polymer Degradation and Stability,* **134**, 76-86.

13. Gu, A., and Liang, G. (2003) Thermal degradation behaviour and kinetic analysis of epoxy/montmorillonite nanocomposites. *Polymer Degradation and Stability,* **80**, 383-391.

14. Pandey, J. K., Reddy, K. R., Kumar, A. P., and Singh, R. P. (2005) An overview on the degradability of polymer nanocomposites. *Polymer Degradation and Stability,* **88**, 234-250.

15. *Polymer-Clay Nanocomposites*, Pinnavaia, T. J., and Beall, G. W. (eds.), John Wiley & Sons Ltd, UK (2000).

16. Celina, M. C., Dayile, A. R., and Quintana, A. (2013) A perspective on the inherent oxidation sensitivity of epoxy materials. *Polymer,* **54**, 3290-3296.

17. Decelle, J., Huet, N., and Bellenger, V. (2003) Oxidation induced shrinkage for thermally aged epoxy networks. *Polymer Degradation and Stability,* **81**, 239-248.

18. Ernault, E., Richaud, E., and Fayolle, B. (2017) Origin of epoxies embrittlement during oxidative ageing. *Polymer Testing,* **63**, 448-454.

19. Zhang, M., Sun, B., and Gu, B. (2016) Accelerated thermal ageing of epoxy resin and 3-D carbon fiber/epoxy braided composites. *Composites, Part A: Applied Science and Manufacturing,* **85**, 163-171.

20. Zhang, M., Zuo, C., Sun, B., and Gu, B. (2016) Thermal ageing degradation mechanisms on compressive behavior of 3-D braided composites in experimental and numerical study. *Composite Structures,* **140**, 180-191.

21. Yang, Y., Xian, G., Li, H., and Sui, L. (2015) Thermal aging of an anhy-

dride-cured epoxy resin. *Polymer Degradation and Stability*, **118**, 111-119.

22. Galant, C., Fayolle, B., Kuntz, M., and Verdu, J. (2010) Thermal and radio-oxidation of epoxy coatings. *Progress in Organic Coatings*, **69**, 322-329.

23. Djouani, F., Zahra, Y., Fayolle, B., Kuntz, M., and Verdu, J. (2013) Degradation of epoxy coatings under gamma irradiation. *Radiation Physics and Chemistry*, **82**, 54-62.

24. Queiroz, D. P. R., Fraïsse, F., Fayolle, B., Kuntz, M., and Verdu, J. (2010) Radiochemical ageing of epoxy coating for nuclear plants. *Radiation Physics and Chemistry*, **79**, 362-364.

25. Paterson-Jones, J. C., Percy, V. A., Giles, R. G. F., and Stephen, A. M. (1973) The thermal degradation of model compounds of amine-cured epoxide resins. II. The thermal degradation of 1,3-diphenoxypropan-2-ol and 1,3-diphenoxypropene. *Journal of Applied Polymer Science*, **17**, 1877-1887.

26. Luda, M. P., Camino, G., Balabanovich, A., I. and Hornung, A. (2002) Scavenging of halogen in recycling of halogen-based polymer materials. *Macromolecular Symposia*, **180**, 141-151, 3.

27. Liu, Y.-L., Hsiue, G.-H., Lan, C.-W., and Chiu, Y.-S. (1997) Phosphorus-containing epoxy for flame retardance: IV. Kinetics and mechanism of thermal degradation. *Polymer Degradation and Stability*, **56**, 291-299.

28. Rosu, D., Cascaval, C. N., and Rosu, L. (2006) Effect of UV radiation on photolysis of epoxy maleate of bisphenol A. *Journal of Photochemistry and Photobiology A: Chemistry*, **177**, 218-224.

29. Decker, C., Keller, L., Zahouily, K., and Benfarhi, S. (2005) Synthesis of nanocomposite polymers by UV-radiation curing. *Polymer*, **46**, 6640-6648.

30. Liu, F., Yin, M., Xiong, B., Zheng, F., Mao, W., Chen, Z., He, C., Zhao, X., and Fang, P. (2014) Evolution of microstructure of epoxy coating during UV degradation progress studied by slow positron annihilation spectroscopy and electrochemical impedance spectroscopy. *Electrochimica Acta*, **133**, 283-293.

31. Modjinou, T., Versace, D.-L., Abbad-Andaloussi, S., Langlois, V., and Renard, E. (2017) Antibacterial and antioxidant photoinitiated epoxy co-networks of resorcinol and eugenol derivatives. *Materials Today Communications*, **12**, 19-28.

32. Burton, B. L. (1993) The thermooxidative stability of cured epoxy resins. I. *Journal of Applied Polymer Science*, **47**, 1821-1837.

33. Ghasemi-Kahrizsangi, A., Shariatpanahi, H., Neshati, J., and Akbarinezhad, E. (2015) Corrosion behavior of modified nano carbon black/epoxy coating in accelerated conditions. *Applied Surface Science*, **331**, 115-126.

34. Wei, Y. H., Zhang, L. X., and Ke, W. (2007) Evaluation of corrosion

protection of carbon black filled fusion-bonded epoxy coatings on mild steel during exposure to a quiescent 3% NaCl solution. *Corrosion Science,* **49**, 287-302.

35. Neisiany, R. E., Khorasani, S. N., Naeimirad, M., Lee, J. K. Y., and Ramakrishna, S. (2017) Improving mechanical properties of carbon/epoxy composite by incorporating functionalized electrospun polyacrylonitrile nanofibers. *Macromolecular Materials and Engineering,* **302**(5), 1600551.

36. Liu M., and Horrocks, A. R. (2002) Effect of carbon black on UV stability of LLDPE films under artificial weathering conditions. *Polymer Degradation and Stability,* **75**, 485-499.

37. Ghasemi-Kahrizsangi, A., Neshati, J., Shariatpanahi, H., and Akbarinezhad, E. (2015) Improving the UV degradation resistance of epoxy coatings using modified carbon black nanoparticles. *Progress in Organic Coatings,* **85**, 199-207.

38. Ghasemi-Kahrizsangi, A., Shariatpanahi, H., Neshati, J., and Akbarinezhad, E. (2015) Degradation of modified carbon black/epoxy nanocomposite coatings under ultraviolet exposure. *Applied Surface Science,* **353**, 530-539.

39. Tcherbi-Narteh, A., Hosur, M., Triggs, E., Owuor, P., and Jelaani, S. (2014) Viscoelastic and thermal properties of full and partially cured DGEBA epoxy resin composites modified with montmorillonite nanoclay exposed to UV radiation. *Polymer Degradation and Stability,* **101**, 81-91.

40. Morlat-Therias, S., Mailhot, B., Gonzalez, D., and Gardette, J.-L. (2005) Photooxidation of polypropylene/montmorillonite nanocomposites. 2. Interactions with antioxidants. *Chemistry of Materials,* **17**, 1072-1078.

41. Tcherbi-Narteh, A., Hosur, M., Triggs, E., and Jeelani, S. (2013) Thermal stability and degradation of diglycidyl ether of bisphenol A epoxy modified with different nanoclays exposed to UV radiation. *Polymer Degradation and Stability,* **98**, 759-770.

42. Lu S.-Y., and Hamerton, I. (2002) Recent developments in the chemistry of halogen-free flame retardant polymers. *Progress in Polymer Science,* **27**, 1661-1712.

43. Zhang, W., He, X., Song, T., Jiao, Q., and Yang, R. (2014) The influence of the phosphorus-based flame retardant on the flame retardancy of the epoxy resins. *Polymer Degradation and Stability,* **109**, 209-217.

44. Qian, X., Song, L., Hu, Y., Yuen, R. K. K., Chen, L., Guo, Y., Hong, N., and Jiang, S. (2011) Combustion and Thermal degradation mechanism of a novel intumescent flame retardant for epoxy acrylate containing phosphorus and nitrogen. *Industrial & Engineering Chemistry Research,* **50**, 1881-1892.

45. Wang, X., Hu, Y., Song, L., Xing, W., Lu, H., Lv, P., *and* Jie, G. (2010)

Flame retardancy and thermal degradation mechanism of epoxy resin composites based on a DOPO substituted organophosphorus oligomer. *Polymer,* **51**, 2435-2445.

46. Jain, P., Choudhary, V., and Varma, I. K. (2003) Phosphorylated epoxy resin: Effect of phosphorus content on the properties of laminates. *Journal of Fire Sciences,* **21**, 5-16.

47. Yang, S., Zhang, Q., and Hu, Y. (2016) Synthesis of a novel flame retardant containing phosphorus, nitrogen and boron and its application in flame-retardant epoxy resin. *Polymer Degradation and Stability,* **133**, 358-366.

48. Yang, S., Wang, J., Huo, S., Cheng, L., and Wang, M. (2015) Preparation and flame retardancy of an intumescent flame-retardant epoxy resin system constructed by multiple flame-retardant compositions containing phosphorus and nitrogen heterocycle. *Polymer Degradation and Stability,* **119**, 251-259.

49. Yang, H., Wang, X., Yu, B., Song, L., Hu, Y., and Yuen, R. K. K. (2012) Effect of borates on thermal degradation and flame retardancy of epoxy resins using polyhedral oligomeric silsesquioxane as a curing agent. *Thermochimica Acta,* **535**, 71-78.

6

Effect of Hygrothermal Ageing on Mechanical, Thermal, Structural and Morphological Properties of Polypropylene and its Blends with Different Copolymers

6.1 Introduction

Polymeric materials are extensively useful for a wide range of industrial, structural, engineering, energy and environmental applications owing to their benefits over other conventional materials. These benefits include satisfactory performance, adjustable properties, light weight structures, low cost, superior corrosion resistance, user-friendly processing, manufacturing and handling, etc. [1-6]. In particular, polypropylene (PP) is a versatile commodity thermoplastic material comprising of many of the above mentioned benefits [7,8]. In order to successfully meet the increasing demand of the polymeric materials with superior property profiles, a large number of physical and chemical approaches for the modification of PP have been developed. Among these approaches, blending PP with other polymeric materials including elastomers, copolymers, thermoplastics, etc., as well as reinforcing PP with functional fillers having nano- or micro-dimensions, have displayed wide acceptance [9-16].

Polymer blending is one of the most common approaches for the development of high performance polymeric materials in order to meet the continuously increasing demands for various functional applications [17-19]. In the case of PP, the blending technique presents importance to acquire enhanced impact performance, which opens opportunities for its use in engineering applications. Accordingly, various copolymers such as styrene-butadiene rubber (SBR) [20], styrene-butadiene-styrene (SBS) [21,22], styrene-ethylene-butylene-styrene (SEBS) [23,24], ethylene-propylene rubber (EPR) [25,26], ethylene-propylene-diene monomer (EPDM) [27,28], ethylene-octene copolymer (EOC) [29-31], ethylene-1-hexene copolymer (EHC) [32] and ethylene-vinyl acetate (EVA) [33] have been

Anish Varghese and Vikas Mittal*, The Petroleum Institute (part of Khalifa University of Science and Technology), Abu Dhabi, UAE
*Current address: Bletchington, Wellington County, Australia

blended with PP for the generation of toughened grades. Many co-polymers also enhance the other physical properties of PP.

The service life of PP based materials in various applications, particularly in outdoor, greatly depends on their performance in hostile environments like humidity, varying temperature and pressure, sunlight, ultraviolet (UV) irradiation, ice or marine conditions, pollutants, chemicals, mechanical loading, biological surroundings, etc., along with their combination [34-36]. Accelerated ageing under simulated hostile environmental conditions in laboratory can be used to predict the material behavior instead of analyzing the long-term natural ageing. Using the accelerated ageing, information related to the life time, maintenance and replacement of the materials can be extracted [37,38]. According to the previous studies, ageing induced changes in the durability of the polymeric materials are related to diminishing mechanical and physical properties owing to the significant structural changes, especially under hygrothermal conditions [39-41]. In addition, the chemical nature of the material as well as time, temperature and medium of immersion present significant influence on the structural changes during hygrothermal ageing [42,43]. Usually, the influence of moisture leads to the detrimental effects such as swelling and degradation of polymers even under normal conditions [6,44-46]. In this respect, accelerated hygrothermal ageing of water immersed samples imitates very aggressive condition which causes moisture diffusion. The introduction of temperature also speeds up such moisture diffusion by following Fickian and/or non-Fickian path [35,44,47-49].

In this study, three sets of PP/copolymer blends in 90/10 and 80/20 weight ratios have been developed. The copolymers chosen to toughen PP by using direct melt blending in a mini twin screw extruder included ethylene vinyl acetate with maleic anhydride grafting (EVA-MA), ethylene butyl acrylate (EBA) and ethylene alpha olefin (EAO). The effect of hygrothermally induced ageing on the properties of neat PP and blend samples was subsequently studied. Simulated hygrothermal ageing conditions were created by immersing the samples in distilled water (DW) and synthetic saline water (SW) at 70 and 90 °C for a period of 32 days. Accordingly, water uptake was measured as a function of hygrothermal ageing time periodically with an interval of 4 days. Specifically, the effect of hygrothermal conditions on impact strength, tensile and thermal properties, structural behavior and morphology of neat PP and PP/copolymer blends was characterized.

6.2 Experimental

6.2.1 Materials

PP homopolymer HD915CF was procured from Abu Dhabi Polymers Company Limited, UAE. The melt flow rate of PP was ca. 8 g/10 min at 230 °C and 2.16 kg load. Also, the reported density value for the PP grade was in the range of 900-910 kg.m^{-3}. Three different commercially procured copolymers included EVA-MA (Fusabond C250, DuPont), EBA (Elvaloy AC 34035, DuPont) and EAO (AFFINITY EG 8200G, Dow Chemical Company). More details of these copolymers are provided in Table 6.1. The materials were used as received from the suppliers.

Table 6.1 Material data for PP and copolymers used in the study

Material	Repeat unit	Appearance	Details
PP homopolymer	$-[CH_2-CH]_n$ with CH_3		Borouge HD915CF **Density:** 900-910 kg.m^{-3} **Melt flow rate:** 8 g/10 min (230 °C, 2.16 kg) **Melting point (MP):** 168.5 °C
Ethylene-vinyl acetate copolymer with maleic anhydride grafting (EVA-MA)	$-[CH_2-CH_2]_n-[CH-CH_2]_m-$ with $O-C(=O)-CH_2$ maleic anhydride group		DuPont Fusabond C250 **Density:** 962 kg. m^{-3} **Melt flow rate:** 1.4 g/10 min (190 °C, 2.16 kg) **MP:** 71 °C
Ethylene-butyl acrylate copolymer (EBA)	$-[CH_2-CH_2]_n-[CH-CH_2]_m-$ with $O=C$, O, CH_2, CH_2, CH_2, CH_3		DuPont Elvaloy AC 34035 **Density:** 930 kg.m^{-3} **Melt flow rate:** 40 g/10 min (190 °C, 2.16 kg) **MP:** 90 °C

Ethylene-alpha olefin copolymer (EAO)	$CH_3-CH_2-CH_2-CH_2$ $-\left[CH_2-CH_2\right]_n-\left[CH-CH_2\right]_m$		**Brand name:** Dow AFFINITY EG 8200G **Density:** 868 kg.m^{-3} **Melt flow rate:** 0.50 g/10 min (190 °C, 2.16 kg) **MP:** 56 °C

6.2.2 Fabrication of PP/Copolymer Blends

Melt mixing in a co-rotating conical twin screw extruder (HAAKE MiniLab, Thermo Scientific, Germany) was performed to fabricate the PP/copolymer blends with weight ratios of 90/10 and 80/20. 5 g batch of the blend samples was generated using 180 °C temperature, 100 rpm screw speed and 15 min blending time. For accurate comparison, neat PP was also subjected to similar processing conditions. Followed by melt extrusion, the samples were injection molded to form specific test specimens, i.e., dumbbells for tensile test and rectangular bars for Izod impact testing, in accordance with the ISO standards with the aid of a lab-scale mini injection molding machine (HAAKE MiniJet PRO, Thermo Scientific, Germany). For injection molding, processing temperature of 180 °C, mold temperature of 125 °C, injection pressure of 430 bar for 10 s and holding pressure of 500 bar for 6 s were used.

6.2.3 Hygrothermal Ageing of Samples

The blend samples were subjected to accelerated hygrothermal ageing conditions to imitate the real life undesirable environmental conditions. Here, the samples were subjected to artificial ageing in distilled water as well as synthetic saline water containing baths at two different temperatures of 70 and 90 °C for a period of 32 days according to the standard ASTM D570-90. Ageing of the samples in DW was useful to imitate the substantial humid environment, which was employed as a reference. SW with 3% sodium chloride concentration was used to examine the ageing behavior under the influence of sea water and to make comparison of the performance of the samples with the DW aged samples.

6.2.4 Characterization

Water Uptake

In order to evaluate the water uptake, the weight of the samples was assessed before ageing (W_0) and periodically after 4 days during ageing (W_t) with the aid of an electronic weighing balance. The periodic weight measurement during ageing was carried out by taking out the samples and wiping the surface water using a dirt-free dry cotton cloth, followed by weighing and re-immersion for the continuation of ageing. An average of five specimens was generally taken for each case. Thus, the weight gain ratio (W_g) of the samples during accelerated ageing was calculated as

$$W_g = \frac{W_t - W_0}{W_0} \times 100$$

where t is the immersion time.

Fourier Transform Infrared (FTIR) Spectroscopy

Structural changes after ageing were analyzed with the aid of FTIR spectrometer BRUKER TENSOR II in transmission mode between 4000 and 400 cm^{-1} wavenumber span with a resolution of 4 cm^{-1}. FTIR spectra of the samples were generated after 32 scans using a diamond attenuated total reflectance (ATR) crystal.

Tensile Properties

In accordance with ISO 527, the tensile properties such as strength, modulus and breaking extension of the samples were assessed using Instron 3345 universal testing machine (UTM). An average of five test specimens (with 75 mm x 5 mm x 2 mm dimensions and 35 mm span length) was recorded. The mechanical performance of the samples was examined at ambient temperature using a crosshead movement rate of 10 mm.min^{-1}.

Izod Impact Strength

In accordance with ISO 180, the Izod impact testing of the unaged and aged samples was performed on a Resil impactor. Un-notched

Izod impact strength values of five rectangular bar shaped test specimens with 80 mm x 10 mm x 4 mm dimensions were averaged for each case. The test was performed at ambient temperature with a hammer speed of 3.64 m.s^{-1}.

Differential Scanning Calorimetry (DSC)

DSC analysis of the unaged and aged samples was performed on a Discovery Series differential scanning calorimeter from TA instruments to evaluate the thermal parameters such as peak melting temperature (T_m), melting enthalpy (ΔH_f), peak crystallization temperature (T_c), crystallization enthalpy (ΔH_c) and percentage crystallinity (X_c). DSC thermograms were generated by scanning ca. 3-8 mg of the samples in dry nitrogen flow (50 mL.min^{-1}). At a rate of 10 °C.min^{-1}, the samples placed in the aluminum pans were scanned by following two sets of heating and cooling cycles within the temperature range of -50 °C to +200 °C, and the second set of analysis was used to generate the test results.

Scanning Electron and Optical Microscopy

The effect of hygrothermal ageing on the fracture micro-morphology of the samples was studied with the aid of a scanning electron microscope (SEM) (FEI Quanta, FEG250, USA) operated using an accelerating voltage of 10 kV. Prior to the microscopy analysis, the impact fractured surface of ca. 3 mm thick test specimens placed on aluminum stubs using conductive carbon tape was sputter coated with gold in order to render them conductive. Ageing induced surface changes in the samples were also examined using an optical microscope (OLYMPUS BX51M, Japan) with an eyepiece magnification of 10x.

Wide Angle X-ray Diffraction (WAXD)

Ageing induced structural changes in the samples were analyzed using an X'Pert PRO Panalytical powder diffractometer by applying CuKα irradiation with a wavelength of 1.5406 Å. The analysis was carried out by employing 40 mA current and 45 kV voltage at room temperature. For the WAXD analysis, small sections of the injection molded samples were scanned using a rate of 0.017° s^{-1} and step time of 10 s in the 2θ angle range of 5-60°.

6.3 Results and Discussion

The water absorption curves (variation of percent water uptake with square root of ageing immersion time) of hygrothermally aged neat PP and its blends with EVA-MA, EBA and EAO in DW and SW are presented in Figure 6.1. The absorption curves of neat PP

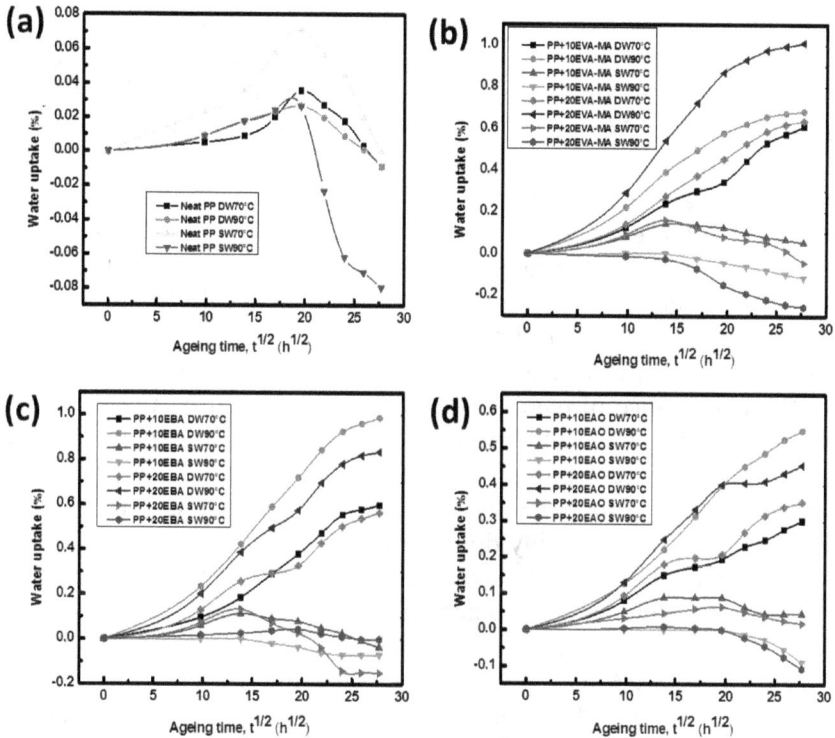

Figure 6.1 Water uptake curves of (a) neat PP, (b) EVA-MA blends, (c) EBA blends and (d) EAO blends.

(Figure 6.1a) indicated two sections, the first section exhibiting the Fickian behavior up to $\sqrt{h}=19.6$ observed by a non-linear water uptake increase with ageing time, whereas the section after $\sqrt{h}=19.6$ indicating the non-Fickian behavior observed by a decreasing water uptake tendency with ageing time. The observed Fickian behavior was related to the solvent diffusion in neat PP, and the optimum % values of water uptake were ca. 0.036 and 0.027 for DW aged samples at 70 °C and 90 °C and ca. 0.071 and 0.027 for respective SW aged samples at 70 °C and 90 °C. Thus, a higher water uptake value for SW aged neat PP at 70 °C was observed among hydrothermally aged PP samples under various conditions. The decreasing water uptake tendency in the second section of the sorption curves of the aged neat PP samples is probably associated with the ageing induced macromolecular chain reorganization, surface crosslinking, washing effect, etc. [38,50,51]. Figures 6.1b, 6.1c and 6.1d illustrate the water uptake behaviors of the respective EVA-MA, EBA and EAO blends under various hygrothermal ageing conditions. The PP/copolymer blends showed almost similar water uptake upon ageing, however, the behavior was different in DW and SW environments. All PP/copolymer blends displayed Fickian behavior in DW at both 70 and 90 °C, which was evidenced by a non-linear increase of water uptake with ageing time (or a combination of initial non-linear increase, succeeded by pseudo-plateau and afterwards non-linear increase). Among the DW aged PP/copolymer blends, 20 wt% EAO blend at 70 °C and 90 °C and 20 wt% EBA blend at 70 °C exhibited combinatorial three stage Fickian behavior. The water uptake increase in DW at extended ageing times is most possibly due to the occurrence of osmotic cracking contributed by the influence of more hydrophilic oligomeric groups in the polymers [52,53]. In addition, the water uptake trend of the PP/copolymer blends in DW was observed to increase with increase in the ageing temperature from 70 to 90 °C probably due to an increase in the free volume resulting from the temperature induced faster rearrangement of the relaxed macromolecular structure [54-56]. Overall, the maximum water uptake values of ca. 1.01, 0.99 and 0.55% were discerned at 90 °C for 20 wt% EVA-MA, 10 wt% EBA and 10 wt% EAO blends respectively. Regardless of the copolymer nature, the copolymer blended PP samples displayed almost similar water sorption behavior upon hygrothermal ageing in SW. At 70 °C, an initial increase in the water uptake followed by a decrease in the weight were noticed. However, at 90 °C, the PP/copolymer blend samples displayed pseudo-plateau behavior nearly close to zero initially, followed by weight decreasing tendency to negative values. In comparison to DW, the observed water diffusion rate was very low in SW due to the influence of salt and the resultant reduction of the chemical potential [51]. Here, the discerned maximum water sorption value of only ca. 0.07% for neat PP supports its persuasive barrier performance which is necessary for protective applications. Compared to PP blends with EVA-MA and EBA, EAO blends displayed higher water uptake resistance, and a maximum value of only 0.55% was noted, whereas the maximum water uptake values of EVA-MA and EBA blends were in the range of 1%. This underlines the superior barrier performance of the EAO toughened PP blends.

To study the structural changes owing to the hygrothermal ageing, FTIR analysis of the unaged and aged samples was carried out in transmission mode, as shown in Figures 6.2 and 6.3. As depicted in

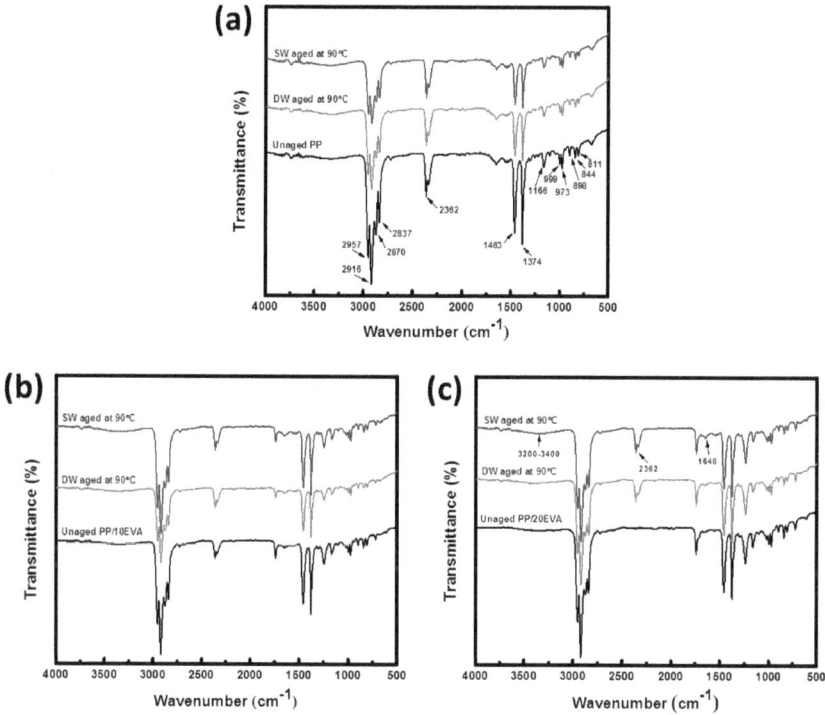

Figure 6.2 FTIR spectra of the unaged and aged samples: (a) neat PP, (b) 10EVA-MA blend and (c) 20EVA-MA blend.

Figure 6.2a, various absorption bands corresponding to CH_3 associated asymmetric and symmetric stretching vibrations at 2957 and 2870 cm^{-1} respectively, CH_2 associated symmetric stretching vibrations at 2916 and 2837 cm^{-1}, C-H associated bending vibration at 2362 cm^{-1}, CH_3 associated asymmetric bending vibration around 1463 cm^{-1}, CH_3 associated symmetric bending vibration around 1374 cm^{-1}, tertiary methyl skeleton associated deformation at 999, 973, 844 and 811 cm^{-1}, stretch direction induced perpendicular absorption nearby 898 cm^{-1}, etc., were observed for neat PP [57]. The

discerned FTIR characteristic bands of EVA-MA, EBA and EAO
blends also presented good resemblance with neat PP more likely

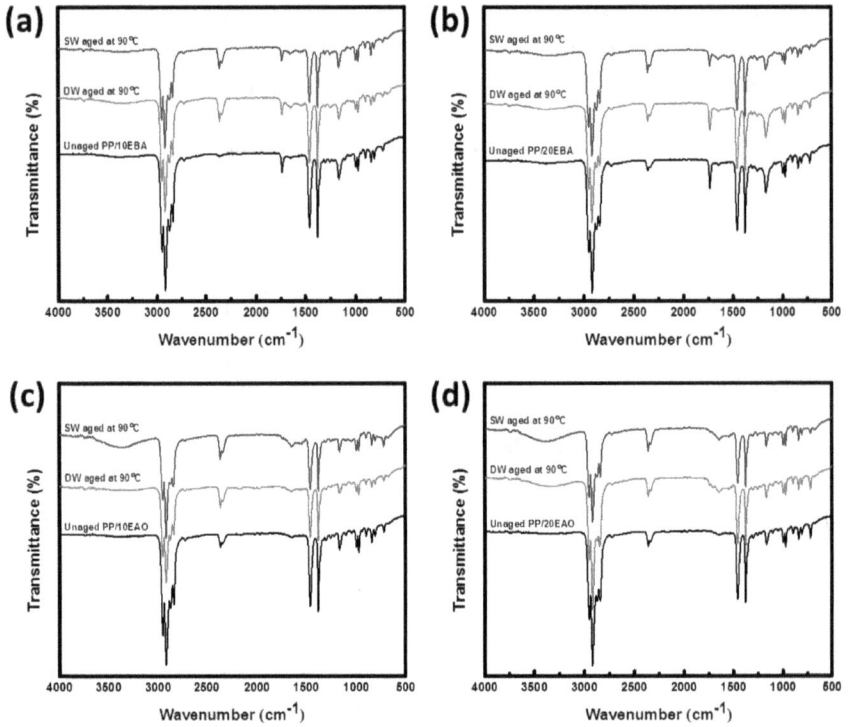

Figure 6.3 FTIR spectra of the unaged and aged samples: (a) 10EBA blend,
(b) 20EBA blend (c) 10EAO blend and (d) 20EAO blend.

due to the compatible blend structures. Following hygrothermal
ageing in both DW and SW, the samples exhibited characteristic IR
bands observed for the unaged specimens, however, the structural
changes could be assessed with the help of intensity variations of
these bands [58,59]. For instance, the characteristic FTIR bands of
the hygrothermally aged neat PP samples exhibited reduced intensity (Figure 6.2a), which is considered to be an indication for the initiation of the ageing induced macromolecular chain scission of neat
PP [55]. For PP blends with EVA-MA, EBA and EAO copolymers, the
discerned intensity reduction was higher for the blends with copolymer content of 10 wt% in comparison with the blends with 20 wt%
copolymer. In addition, the FTIR spectra of the aged PP/copolymer
blend samples displayed enhanced intensity for the absorption

bands corresponding to absorbed water associated asymmetric bending in the broad range of 3200-3400 cm^{-1}, water induced C-H bending vibration nearby 2362 cm^{-1} and water induced OH stretching nearby 1640 cm^{-1} [47,58].

The WAXD plots of the samples, shown in Figures 6.4 and 6.5, provide insights into the hydrothermal ageing induced crystallinity variations. As depicted in Figure 6.4a, the unaged neat PP displayed

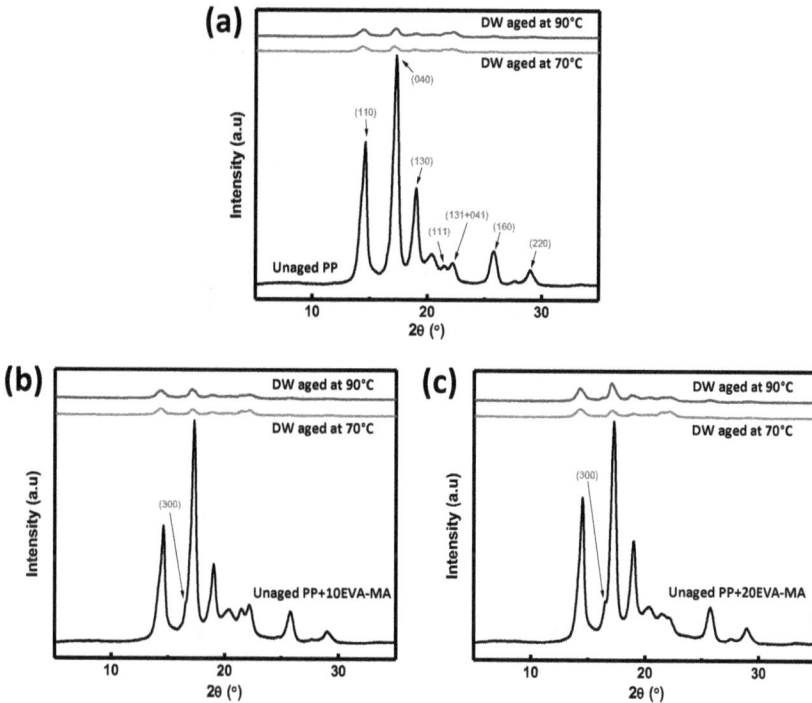

Figure 6.4 WAXD patterns of unaged and aged samples: (a) neat PP, (b) 10EVA-MA blend and (c) 20EVA-MA blend.

seven characteristic diffraction peaks associated with α crystalline planes (110) at 14.55°, (040) at 17.26°, (130) at 19.03°, (111) at 21.47°, (131+041) at 22.18°, (160) at 25.76° and (220) at 29.01° [60-62]. Also, the inter-planar distance (*d*) corresponding to these peaks was estimated using the Bragg's relation, $d = \lambda/(2\sin\theta_{max})$, where λ implies the X-ray wavelength (1.5406 Å) applied for the analysis. Hence, the estimated inter-planar distance values were 6.08 Å for (110), 5.13 Å for (040), 4.66 Å for (130), 4.14 Å for (111), 4.00Å for (131+041), 3.46Å for (160) and 3.08 Å for (220). After hygrothermal ageing in DW at 70 and 90 °C, the appearance of same characteristic crystalline peaks was observed, differing only in intensity. In actual, the intensity of all crystalline peaks was observed

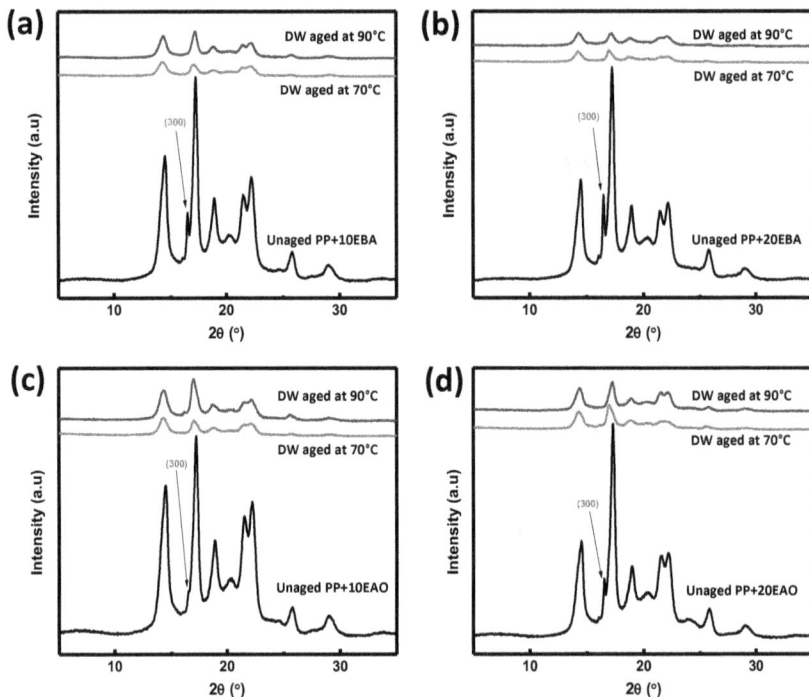

Figure 6.5 WAXD patterns of unaged and aged samples: (a) 10EBA blend, (b) 20EBA blend (c) 10EAO blend and (d) 20EAO blend.

to considerably decrease for aged neat PP, owing to the ageing induced macromolecular chain scission and corresponding crystallinity reduction [44]. The observed intensity reduction was higher for the neat PP aged in DW at 70 °C in comparison to 90 °C, suggesting the ageing induced macromolecular reorganization at 90° C. The WAXD plots of EVA-MA, EBA and EAO blends revealed the appearance of same α crystalline peaks as neat PP at the same diffraction angles, together with an additional β crystalline peak with a plane of (300) at 16.53° and the corresponding inter-planar distance of 5.36 Å. For PP/copolymer blends, similar crystalline peak intensity reduction was observed after hygrothermal ageing in DW at 70 and 90 °C. Among the blends, the crystalline peak intensity reduction was observed to be lower for EAO blends.

For a material, impact strength indicates its ability to absorb sudden force, thus, resulting in effective plastic deformation without any fracture development. Table 6.2 compares the impact strength

Table 6.2 Mechanical properties of unaged and aged samples

Sample	Impact strength (KJ/m²)	Tensile strength (MPa)	Tensile modulus (MPa)	Breaking extension (%)
Unaged PP	7	39	1011	24
DW aged PP at 70 °C	51	36	815	23
DW aged PP at 90 °C	46	35	808	21
SW aged PP at 70 °C	45	36	809	21
SW aged PP at 90 °C	46	35	801	21
10EVA-MA	26	33	1095	38
DW aged 10EVA-MA at 70 °C	32	31	837	30
DW aged 10EVA-MA at 90 °C	33	30	764	29
SW aged 10EVA-MA at 70 °C	29	29	720	26
SW aged 10EVA-MA at 90 °C	39	28	705	26
20EVA-MA	42	30	1143	43
DW aged 20EVA-MA at 70 °C	35	27	718	28
DW aged 20EVA-MA at 90 °C	38	26	659	26
SW aged 20EVA-MA at 70 °C	31	27	710	26
SW aged 20EVA-MA at 90 °C	41	26	651	25
10EBA	54	32	1115	44
DW aged 10EBA at 70 °C	60	31	767	34
DW aged 10EBA at 90 °C	61	31	684	32
SW aged 10EBA at 70 °C	58	31	756	31
SW aged 10EBA at 90 °C	60	31	681	30
20EBA	74	26	1036	299
DW aged 20EBA at 70 °C	79	26	567	81
DW aged 20EBA at 90 °C	78	25	495	80
SW aged 20EBA at 70 °C	78	26	566	77
SW aged 20EBA at 90 °C	77	25	490	61
10EAO	19	35	1244	28
DW aged 10EAO at 70 °C	74	34	837	27
DW aged 10EAO at 90 °C	70	32	767	26
SW aged 10EAO at 70 °C	52	32	710	26
SW aged 10EAO at 90 °C	100	29	628	25
20EAO	96	30	1084	263
DW aged 20EAO at 70 °C	101	28	673	47
DW aged 20EAO at 90 °C	88	28	627	56
SW aged 20EAO at 70 °C	98	28	664	43
SW aged 20EAO at 90 °C	59	28	620	32

of the samples before and after ageing in both DW and SW at 70 and 90 °C. It can be observed that the addition of 10 and 20 wt% of co-polymers was useful in appreciably improving the impact strength of neat PP. In the PP/copolymer blends, the copolymeric dispersed phase is expected to help in generating micro-voids in the matrix by undergoing crazing as well as shear yielding through effective ab-sorption of the majority of the impact energy by acting as stress ac-cumulator. Thus, the resultant PP/copolymer blends show favorable dissipation of the impact energy by undergoing fracture-free plastic deformation [31,63,64]. The impact strength of neat PP was ob-served to increase significantly in all hygrothermal ageing condi-tions by virtue of the water induced plasticization effect [41,65]. Therefore, the appearance of a highly appreciable impact strength retention rate of more than 550% was noted for the aged neat PP samples (Figure 6.6). Hygrothermally aged PP/copolymer blends also exhibited an enhancement of the impact strength except PP

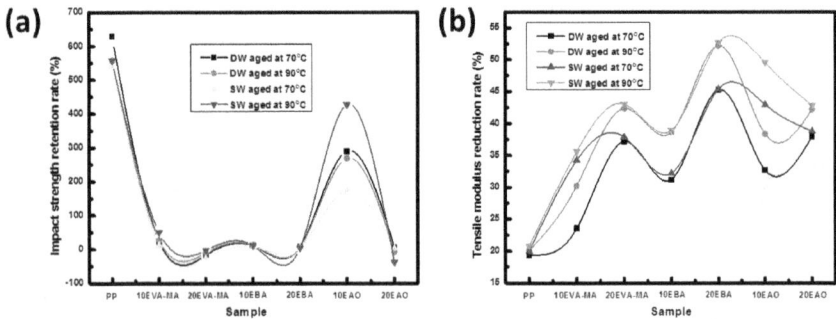

Figure 6.6 Variation of (a) impact strength retention rate and (b) tensile modulus reduction rate of neat PP, EVA-MA blends, EBA blends and EAO blends in various hygrothermal conditions.

blends with 20 wt% EVA-MA in all ageing conditions and 20 wt% EAO in both DW and SW at 90 °C (Figure 6.6). The drop in the im-pact strength of such PP/copolymer blends is most probably related to the structural integrity loss in hygrothermal conditions [44]. Among various PP/copolymer blend systems, the blend with 10 wt% EAO displayed higher rate of impact strength retention, ob-served in the range of 174-426% (Table 6.2). The observed impact strength values also revealed superior impact strength retention rate for 10 wt% copolymer blends than the blends with 20 wt% co-polymer in hygrothermal conditions. In addition, the impact strength retention capability of a majority of the PP/copolymer blend samples was observed to increase with temperature in both DW and SW. Such increasing tendency of the impact strength retention rate at higher temperature relates to the increased degree of water induced plasticization effect at the blend interface and, thus, easier generation of micro-voids [41,44].

The mechanical properties such as strength, modulus and breaking extension of unaged and hygrothermally aged neat PP, EVA-MA blends, EBA blends and EAO blends are presented in Table 6.2. An improvement in the tensile modulus of PP was observed after blending with copolymers. The observed behavior indicates the ability of the blend constituents to proceed with stretching induced deformation simultaneously in accordance with the generation of defect-free interface owing to good phase compatibility [29]. On comparing the effect of copolymers, 20 wt% EVA-MA and 10 wt% EBA and EAO blended PP were observed to exhibit higher improvement in tensile modulus. The tensile modulus of the neat PP and PP/copolymer blends was observed to be negatively affected under hygrothermal ageing conditions. The modulus deteriorated proportionately with ageing temperature in both DW and SW. At the same time, the extent of tensile modulus deterioration was observed to be higher in SW as compared to DW. For instance, neat PP showed ca. 19-20% deterioration in DW and ca. 20-21% deterioration in SW (Figure 6.6). In the case of PP/copolymer blends, the tensile modulus deterioration rate was observed to increase with increase in the copolymer content. Overall, 10 wt% EVA-MA blend exhibited a lower rate of tensile modulus deterioration of ca. 24-30% and 34-36% in DW and SW respectively as compared to other blends. On the other hand, the observed tensile modulus deterioration in respective DW and SW were 37-42% and 38-43% for 20 wt% EVA-MA blend, 31-39% and 32-39% for 10 wt% EBA blend, 45-52% and 45-53% for 20 wt% EBA blend, 33-38% and 43-50% for 10 wt% EAO blend and 38-42% and 39-43% for 20 wt% EAO blend (Figure 6.6). Such deterioration in the tensile modulus might be due to the reduced material resistance resulting from the macromolecular breakdown induced by the absorbed water in hygrothermal conditions. Also, the extent of such structural changes was observed to increase with the ageing temperature as a result of increased water penetration tendency [55,66]. Thus, the immersion in SW caused higher tensile modulus deterioration as compared to DW due to the development of more micro-cracks in high pH environment [34,67].

The tensile strength of PP was observed to reduce after blending with copolymers and the extent of reduction was noticed to increase with the copolymer content. The observed reduction tendency after copolymer addition is possibly caused by the deterioration in load-carrying area. Similar to tensile modulus, hygrothermal ageing impacted the tensile strength negatively (Table 6.2). Overall, among the samples, the EBA blends exhibited superior resistance against the ageing induced tensile strength reduction under hygrothermal conditions. In addition, the elongation at break of neat PP was observed to enhance after copolymer addition. Upon tensile stretching, the dispersed copolymeric phase in the blend system can serve as effective stress absorber, thus, leading to successful macromolecular extension. The extension at break of the samples also deteriorated under hygrothermal conditions. Compared to neat PP, the observed breaking extension deterioration rate was higher for the PP/copolymer blends, which underlines the contribution of the amorphous portion for water penetration and macromolecular breakdown, as observed previously [58].

Table 6.3 presents the DSC data obtained from second heating and cooling scans shown in Figures 6.7-6.10 for unaged and hygrothermally aged samples. In order to estimate the X_c of the unaged and aged samples, the following equation was used [52,68]:

$$X_c = \frac{\Delta H}{\Delta H_0 w_{pp}} \times 100$$

In this equation, the parameters ΔH_0, ΔH and w_{pp} represent theoretical melting enthalpy of fully crystalline PP (207.1 J/g) [68], DSC generated melting enthalpy of the studied samples and weight fraction of PP in the studied samples, respectively.

As observed from Table 6.3, the unaged neat PP exhibited the peak melting temperature of 168.5 °C. Also, the T_m values within a standard deviation of -1 °C were discerned after blending PP with 10 and 20 wt% of EVA-MA, EBA and EAO, which possibly indicates the blends with defect-free PP crystal structure [69]. After exposure to the accelerated hygrothermal ageing conditions, T_m of neat PP displayed slight reduction in DW, whereas it remained almost same in SW. A similar trend was observed for the PP blends with 10 wt% EVA-MA and EBA, though virtually similar T_m was observed for PP blends with 20 wt% EVA-MA and EBA. In the case of EAO blends, T_m remained same or slightly higher after hygrothermal exposure. In

Table 6.3 DSC data for unaged and aged samples (T_c: crystallization temperature; ΔH_c: crystallization enthalpy; Tm: melting temperature; ΔH_f: melting enthalpy; X_c: percentage crystallinity; ΔX_c: percentage crystallinity variation)

Sample	T_c (°C)	ΔH_c (J/g)	T_m (°C)	ΔH_f (J/g)	X_c (%)	ΔX_c (%)
Unaged PP	127.9	98.7	168.5	90.8	43.8	-
DW aged PP at 70 °C	126.3	106.2	166.9	112.2	54.2	10.4
DW aged PP at 90 °C	126.4	110.1	166.9	108.0	52.1	8.3
SW aged PP at 70 °C	126.2	111.2	168.2	106.5	51.4	7.6
SW aged PP at 90 °C	126.2	107.0	168.6	105.4	50.9	7.1
Unaged 10EVA-MA	127.9	90.5	168.3	87.3	46.8	-
DW aged 10EVA-MA (70 °C)	126.6	102.3	167.7	100.8	54.1	7.3
DW aged 10EVA-MA (90 °C)	126.6	98.3	167.7	96.8	51.9	5.1
SW aged 10EVA-MA (70 °C)	125.8	99.3	168.1	96.7	51.9	5.1
SW aged 10EVA-MA (90 °C)	125.8	95.7	168.5	93.7	50.3	3.5
Unaged 20EVA-MA	128.2	81.0	167.7	80.7	48.7	-
DW aged 20EVA-MA (70 °C)	125.0	86.9	167.7	88.5	53.5	4.8
DW aged 20EVA-MA (90 °C)	124.6	88.8	168.1	88.2	53.2	4.5
SW aged 20EVA-MA (70 °C)	125.0	84.0	169.2	81.8	49.4	0.7
SW aged 20EVA-MA (90 °C)	125.5	88.2	168.8	85.7	51.7	3.0
Unaged 10EBA	127.8	89.7	168.5	83.1	44.6	-
DW aged 10EBA (70 °C)	126.2	100.3	167.7	98.3	52.7	8.1
DW aged 10EBA (90 °C)	126.4	97.4	167.7	94.0	50.0	5.4
SW aged 10EBA (70 °C)	125.8	90.2	168.6	84.4	45.3	0.7
SW aged 10EBA (90 °C)	126.2	97.1	168.2	92.2	49.5	4.9
Unaged 20EBA	127.9	80.9	168.1	79.1	47.8	-
DW aged 20EBA (70 °C)	125.8	86.1	168.5	87.6	52.9	5.1
DW aged 20EBA (90 °C)	126.6	88.8	167.8	86.6	52.3	4.5
SW aged 20EBA (70 °C)	125.8	85.7	168.5	83.6	50.5	2.7
SW aged 20EBA (90 °C)	125.4	86.9	168.5	83.6	50.5	2.7
Unaged 10EAO	127.8	91.4	167.7	87.9	47.2	-
DW aged 10EAO (70 °C)	126.0	100.7	168.5	99.5	53.4	6.2
DW aged 10EAO (90 °C)	126.7	98.7	167.7	96.5	51.8	4.6
SW aged 10EAO (70 °C)	125.6	93.3	168.5	91.4	49.0	1.8
SW aged 10EAO (90 °C)	126.2	91.8	168.5	89.9	48.2	1.0
Unaged 20EAO	128.2	82.6	167.7	78.6	47.5	-
DW aged 20EAO (70 °C)	126.6	87.7	167.7	88.2	53.2	5.7
DW aged 20EAO (90 °C)	125.8	82.5	168.1	81.2	48.0	0.5
SW aged 20EAO (70 °C)	125.6	84.5	168.5	81.4	49.1	1.6
SW aged 20EAO (90 °C)	125.4	88.0	168.8	85.1	51.9	4.4

addition, the samples aged in SW displayed virtually same or slightly higher value of T_m in comparison to the DW aged samples exhibiting a slightly reduced T_m in majority of the cases (except EAO based blends). From these observations, the capability of the EAO based blend samples to withstand the hygrothermal ageing induced T_m

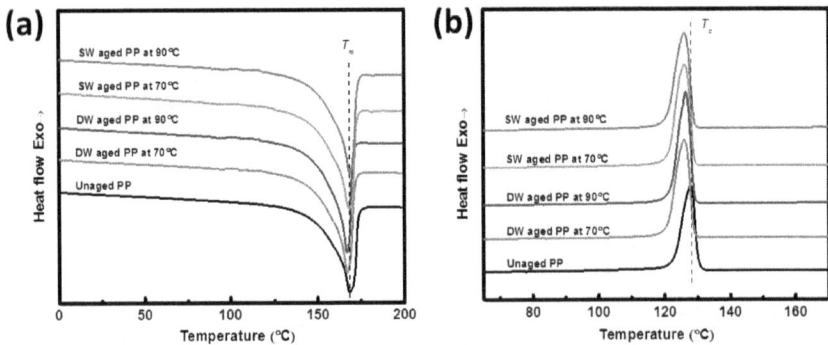

Figure 6.7 DSC (a) heating and (b) cooling curves of unaged and hygrothermally aged neat PP samples.

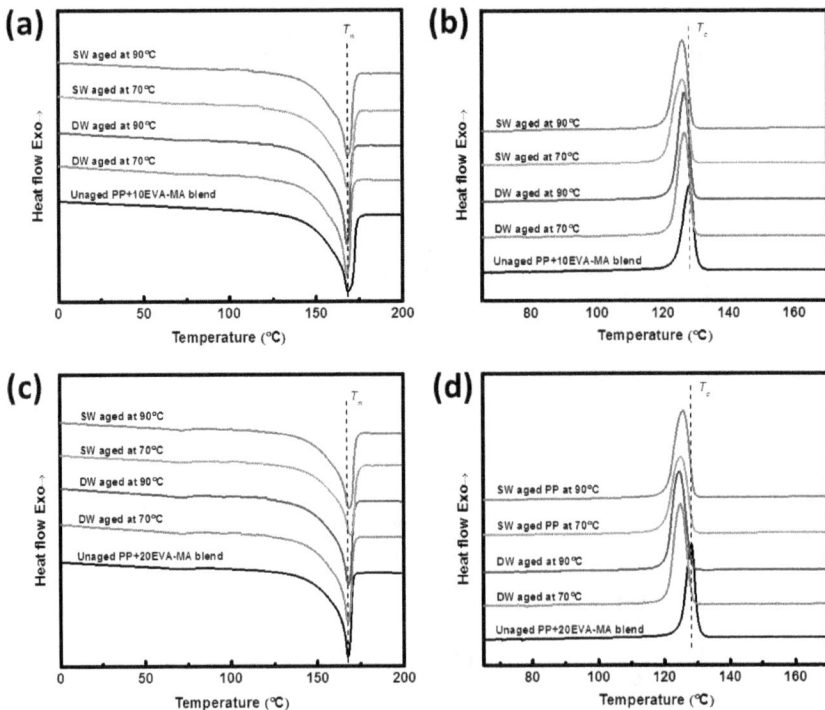

Figure 6.8 DSC heating and cooling curves of unaged and aged samples: (a,b) 10EVA-MA blend and (c,d) 20EVA-MA blend.

reduction in both DW and SW at 70 and 90 °C can be confirmed. This is in good agreement with the observations related to the water uptake discussed earlier.

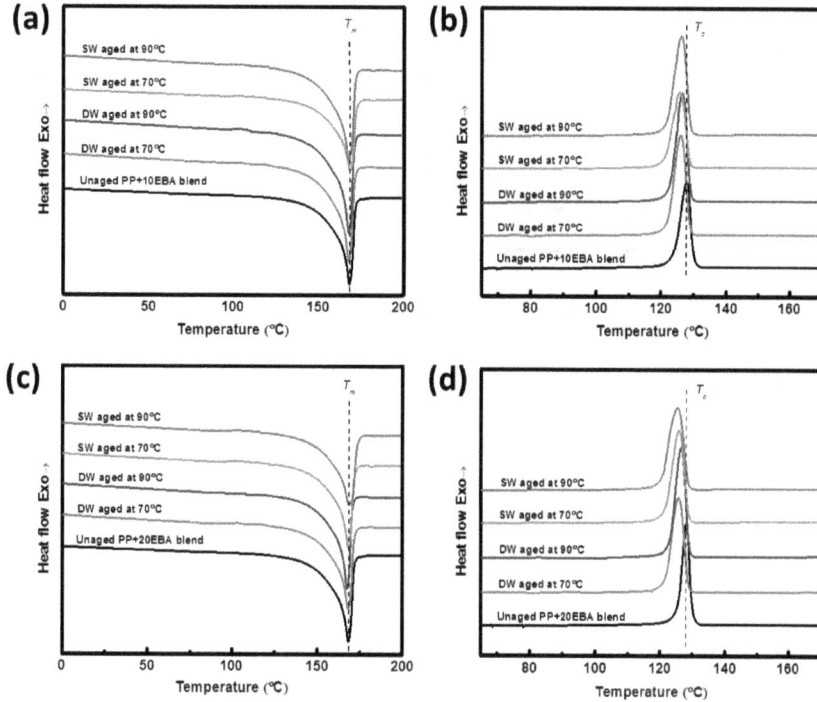

Figure 6.9 DSC heating and cooling curves of unaged and aged samples: (a,b) 10EBA blend and (c,d) 20EBA blend.

Concerning T_c, the unaged neat PP displayed a value of 127.9 °C and the blends with 10 and 20 wt% EVA-MA, EBA and EAO displayed virtually similar values. It indicated that the arrangement of PP crystallites remained undisturbed in the presence of copolymers [64]. The T_c of hygrothermally aged neat PP and PP/copolymer blends was observed to decrease, due to the influence of moisture, thus, altering the crystal structure perfection.

The observed X_c for the unaged neat PP was 43.8%, which was found to increase after blending with copolymers owing to a probable copolymer mediated ordering of PP crystallites [70]. X_c was also observed to increase for all samples after hygrothermal ageing (Table 6.3). For instance, the aged neat PP exhibited an increase of crystallinity in the range of 7.1-10.4% under various ageing conditions.

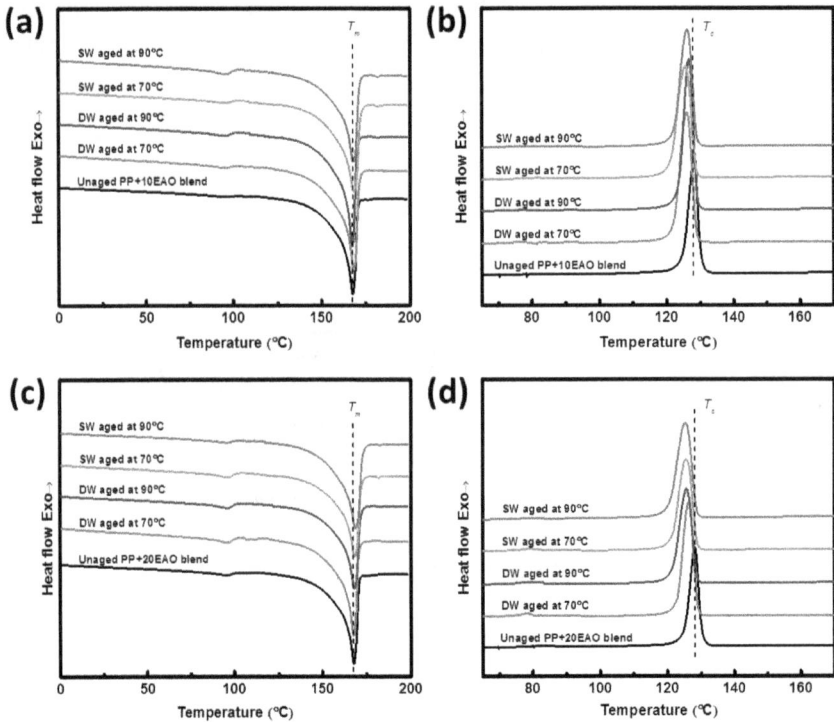

Figure 6.10 DSC heating and cooling curves of unaged and aged samples: (a,b) 10EAO blend and (c,d) 20EAO blend.

In addition, the observed ΔX_c of the blend samples was in the range of 3.5-7.3%, 0.7-4.8%, 0.7-8.1%, 2.7-5.1%, 1.0-6.2% and 0.5-5.7% for the blends with 10 wt% EVA-MA, 20 wt% EVA-MA, 10 wt% EBA, 20 wt% EBA, 10 wt% EAO and 20 wt% EAO respectively. Indeed, the noticed positive ΔX_c trend was lower for the PP blends with 20 wt% copolymers as compared to the PP blends with 10 wt% copolymer content. Also, the samples aged in DW exhibited higher enhancement in X_c as compared to SW, which also correlates with the observations pertaining to water uptake. Among various blends, the EAO blends demonstrated a lower degree of X_c increase, which indicated some degree of resistance against the ageing induced macromolecular chain scission under hygrothermal conditions.

The micro-morphology of the unaged and aged samples was studied using SEM of the gold sputter-coated impact fractured surfaces, as presented in Figures 6.11-6.13. Comparing the images of the unaged PP and PP/copolymer blend samples, a large number of

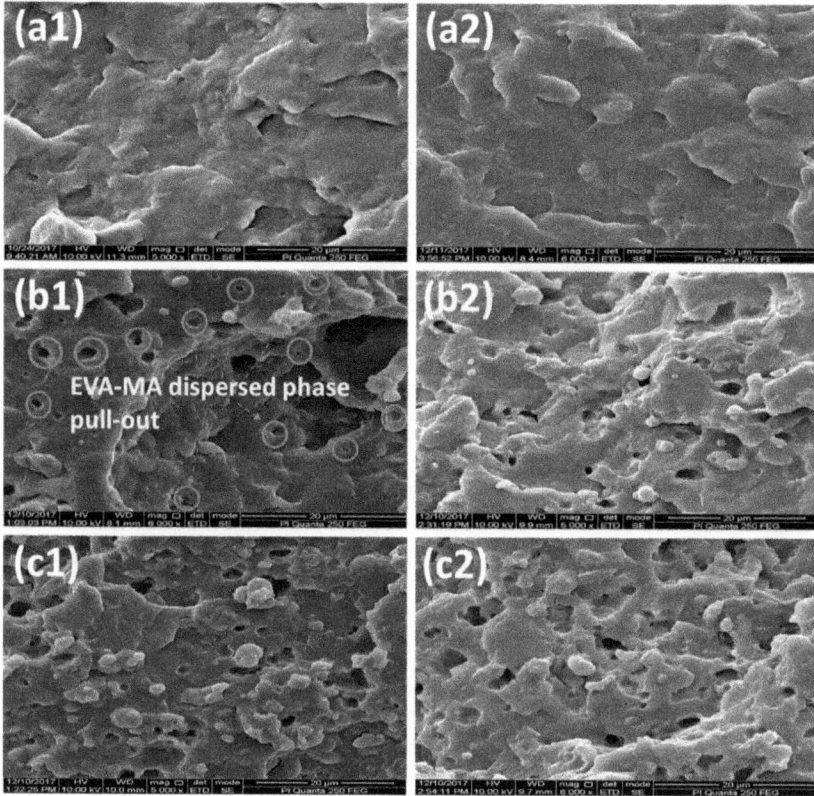

Figure 6.11 SEM images of the impact fractured surface for unaged and aged samples of ((a1) and (a2)) neat PP, ((b1) and (b2)) 10EVA-MA blend and ((c1) and (c2)) 20EVA-MA blend respectively.

spherical shaped dispersed copolymer phase pull-outs, also named as micro-voids, were observed in the PP matrix in the PP/copolymer blends, which correspondingly resulted in improved impact strength. This observation underlines the successful debonding or cavitation at the blend interface upon impact loading, by absorbing the majority of the force and consequent generation of crazing and shear yielding. From the SEM fractographs of the unaged PP/copolymer blends, uniform copolymeric phase distribution could also be observed in the continuous PP phase. Indeed, on increasing the copolymer content, the extent of shock absorbing sites was observed to increase, along with enhanced dispersed copolymeric phase size. It could be observed from Figures 6.11a1 and 6.11a2 that the fracture morphology changed from brittle to ductile

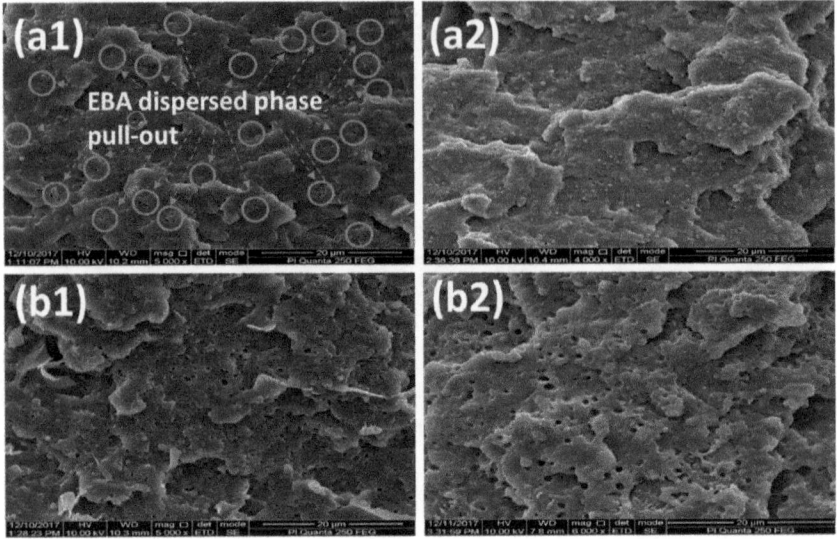

Figure 6.12 SEM images of the impact fractured surface for unaged and aged samples of ((a1) and (a2)) 10EBA blend and ((b1) and (b2)) 20EBA blend.

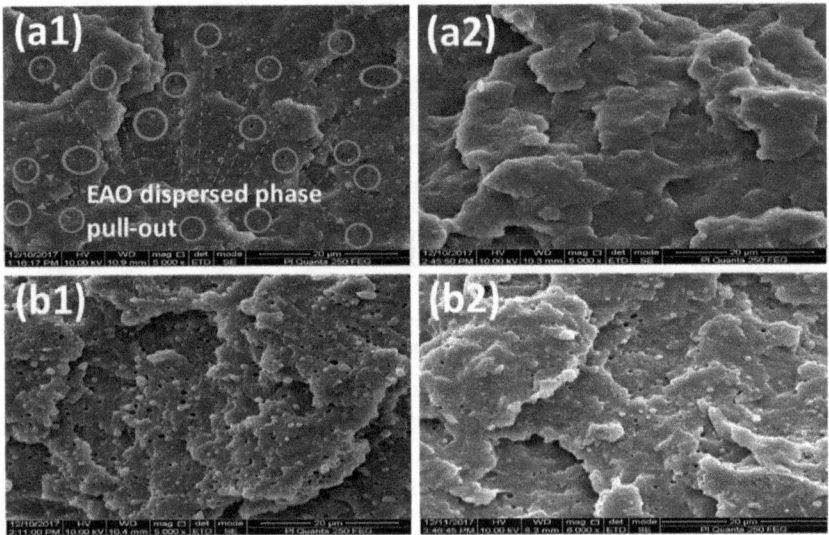

Figure 6.13 SEM images of the impact fractured surface for unaged and aged samples of ((a1) and (a2)) 10EAO blend and ((b1) and (b2)) 20EAO blend.

for neat PP after ageing (90 °C DW) due to the influence of the water induced plasticizing effect, which accordingly resulted in 557% enhancement in the impact strength. For the 90 °C DW aged 10 wt% copolymer blends (Figures 6.11b2, 6.12a2 and 6.13a2), the impact strength was also observed to exhibit an increasing trend due to the influence of the ageing induced water plasticizing effect, along with an improvement in the interfacial characteristics. Among the 10 wt% copolymer toughened PP samples, EAO blend displayed an optimal enhancement of ca. 268%. Considering 20 wt% copolymer toughened PP samples, ageing had no significant effect on the impact strength, as EVA-MA and EAO containing blends displayed a small decrease whereas EBA blend showed a small increase. This is due to the simultaneous presence of the ageing induced water plasticizing effect and decohesion of the copolymeric dispersed phase, as displayed in Figures 6.11c2, 6.12b2 and 6.13b2.

Figure 6.14 displays the optical micrographs of the 90 °C DW

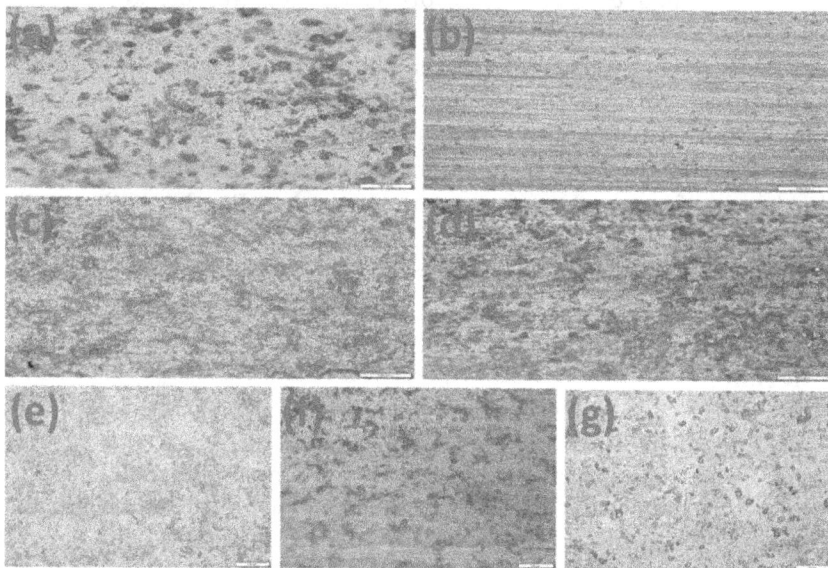

Figure 6.14 Surface micrographs of aged (a) neat PP, (b) 10EVA-MA blend, (c) 20EVA-MA blend, (d) 10EBA blend, (e) 20EBA blend, (f) 10EAO blend and (g) 10EAO blend.

aged samples. Surface deterioration trend through the loss of surface uniformity was observed in the samples. Also, the surface roughness of the aged samples was observed to increase, evidenced by the appearance of irregular pore like structures and small cracks on the surface. As highlighted in Figure 6.14a, neat PP experienced higher surface deterioration tendency compared to the PP/copolymer blend samples in hygrothermal ageing conditions. Overall, the PP blends with lower content (10 wt%) of EVA-MA and higher content (20 wt%) of EBA and EAO displayed superior resistance against surface deterioration.

6.4 Conclusions

In this study, PP/copolymer blends using EVA-MA, EBA and EAO copolymers in 10 and 20 wt% compositions were generated by direct melt blending in a twin screw extruder. Water uptake of the samples as a function of the hygrothermal ageing time exhibited a higher water uptake trend in DW as compared to SW. The impact strength was observed to increase due to the water plasticizing effect and improved interfacial properties in the blend systems, whereas the tensile strength, modulus and extension at break exhibited a decreasing trend due to the reduced material resistance resulting from the macromolecular breakdown caused by the combined effect of water and thermal ageing. The observations related to the impact strength were further supported by SEM analysis of the impact fractured surface of the samples. For hygrothermally aged samples, the DSC generated normalized X_c was also observed to increase by virtue of chemi-crystallization phenomenon.

References

1. Taktak, R., Guermazi, N., Derbeli, J., and Haddar, N. (2015) Effect of hygrothermal aging on the mechanical properties and ductile fracture of polyamide 6: Experimental and numerical approaches. *Engineering Fracture Mechanics*, **148**, 122-133.
2. Deroiné, M., Le Duigou, A., Corre, Y.-M., Le Gac, P.-Y., Davies, P., César, G., and Bruzaud, S. (2014). Accelerated ageing of polylactide in aqueous environments: comparative study between distilled water and seawater. *Polymer Degradation and Stability*, **108**, 319-329.
3. Eftekhari, M., and Fatemi, A. (2016) Tensile behavior of thermoplastic composites including temperature, moisture, and hygrothermal effects. *Polymer Testing*, **51**, 151-164.
4. Deroiné, M., Le Duigou, A., Corre, Y.-M., Le Gac, P.-Y., Davies, P.,

César, G., and Bruzaud, S. (2014) Accelerated ageing and lifetime prediction of poly (3-hydroxybutyrate-co-3-hydroxyvalerate) in distilled water. *Polymer Testing*, **39**, 70-78.

5. *Thermally STable and Flame Retardant Polymer Nanocomposites*, Mittal, V. (ed.), Cambridge University Press, UK (2011).

6. Wang, M., Xu, X., Ji, J., Yang, Y., Shen, J., and Ye, M. (2016) The hygrothermal aging process and mechanism of the novolac epoxy resin. *Composites, Part B: Engineering*, **107**, 1-8.

7. Bikiaris, D., Vassiliou, A., Chrissafis, K., Paraskevopoulos, K., Jannakoudakis, A., and Docoslis, A. (2008) Effect of acid treated multiwalled carbon nanotubes on the mechanical, permeability, thermal properties and thermo-oxidative stability of isotactic polypropylene. *Polymer Degradation and Stability*, **93**, 952-967.

8. Wang, Y., and Tsai, H. B. (2012) Thermal, dynamic-mechanical, and dielectric properties of surfactant intercalated graphite oxide filled maleated polypropylene nanocomposites. *Journal of Applied Polymer Science*, **123**, 3154-3163.

9. Xu, J., Mittal, V., and Bates, F. S. (2016) Toughened isotactic polypropylene: phase behavior and mechanical properties of blends with strategically designed random copolymer modifiers. *Macromolecules*, **49**, 6497-6506.

10. Ma, L.-F., Bao, R.-Y., Dou, R., Zheng, S.-D., Liu, Z.-Y., Zhang, R.-Y., Yang, M.-B., and Yang, W. (2016) Conductive thermoplastic vulcanizates (TPVs) based on polypropylene (PP)/ethylene-propylene-diene rubber (EPDM) blend: From strain sensor to highly stretchable conductor. *Composites Science and Technology*, **128**, 176-184.

11. Gao, Y., Li, J., Li, Y., Yuan, Y., Huang, S., and Du, B. (2017) Effect of Elastomer Type on Electrical and Mechanical Properties of Polypropylene/Elastomer Blends. *Electrical Insulating Materials (ISEIM) Symposium*, pp. 574-577.

12. Zhou, Y., He, J., Hu, J., Huang, X., and Jiang, P. (2015) Evaluation of polypropylene/polyolefin elastomer blends for potential recyclable HVDC cable insulation applications. *IEEE Transactions on Dielectrics and Electrical Insulation*, **22**, 673-681.

13. Yang, C.-j., Huang, T., Yang, J.-h., Zhang, N., Wang, Y., and Zhou, Z.-w. (2017) Carbon nanotubes induced brittle-ductile transition behavior of the polypropylene/ethylene-propylene-diene terpolymer blends. *Composites Science and Technology*, **139**, 109-116.

14. Hong, C. H., Lee, Y. B., Bae, J. W., Jho, J. Y., Nam, B. U., and Hwang, T. W. (2005) Preparation and mechanical properties of polypropylene/clay nanocomposites for automotive parts application. *Journal of Applied Polymer Science*, **98**, 427-433.

15. Zhang, X., Yan, X., He, Q., Wei, H., Long, J., Guo, J., Gu, H., Yu, J., Liu, J., and Ding, D. (2015) Electrically conductive polypropylene nanocomposites with negative permittivity at low carbon nanotube loa-

ding levels. *ACS Applied Materials and Interfaces*, **7**, 6125-6138.

16. Qiu, F., Hao, Y., Li, X., Wang, B., and Wang, M. (2015) Functionalized graphene sheets filled isotactic polypropylene nanocomposites. *Composites, Part B: Engineering*, **71**, 175-183.

17. *Functional Polymer Blends: Synthesis, Properties and Performance*, Mittal, V. (ed.), CRC Press, USA (2012).

18. *Characterization of Polymer Blends*, Thomas, S., Grohens, Y., and Jyotishkumar, P. (eds.), WIly VCH, Germany (2014).

19. Ibrahim, B. A., and Kadum, K. M. (2010) Influence of polymer blending on mechanical and thermal properties. *Modern Applied Science*, **4**, 157.

20. Cook, R. F., Koester, K. J., Macosko, C. W., and Ajbani, M. (2005) Rheological and mechanical behavior of blends of styrene-butadiene rubber with polypropylene. *Polymer Engineering and Science*, **45**, 1487-1497.

21. Abreu, F., Forte, M., and Liberman, S. (2005) SBS and SEBS block copolymers as impact modifiers for polypropylene compounds. *Journal of Applied Polymer Science*, **95**, 254-263.

22. Das, V., Gautam, S. S., and Pandey, A. K. (2011) Effect of SBS content on low temperature impact strength, morphology and rheology of PP-cp/SBS blends. *Polymer-Plastics Technology and Engineering*, **50**, 825-832.

23. Balkan, O., Demirer, H., and Kayali, E. S. (2011) Effects of deformation rates on mechanical properties of PP/SEBS blends. *Journal of Achievements in Materials and Manufacturing Engineering*, **47**, 26-33.

24. Mae, H., Omiya, M., and Kishimoto, K. (2008) Material ductility and toughening mechanism of polypropylene blended with bimodal distributed particle size of styrene–ethylene–butadiene–styrene triblock copolymer at high strain rate. *Journal of Applied Polymer Science*, **110**, 3941-3953.

25. Zebarjad, S., Bagheri, R., Reihani, S., and Lazzeri, A. (2003) Deformation, yield and fracture of elastomer-modified polypropylene. *Journal of Applied Polymer Science*, **90**, 3767-3779.

26. Kim, M. H., Alamo, R. G., and Lin, J. (1999) The cocrystallization behavior of binary blends of isotactic polypropylene and propylene-ethylene random copolymers. *Polymer Engineering and Science*, **39**, 2117-2131.

27. Balaji, A. B., Ratnam, C. T., Khalid, M., and Walvekar, R. (2017) Effect of electron beam irradiation on thermal and crystallization behavior of PP/EPDM blend. *Radiation Physics and Chemistry*, **141**, 179-189.

28. Bouchart, V., Bhatnagar, N., Brieu, M., Ghosh, A., and Kondo, D. (2008) Study of EPDM/PP polymeric blends: mechanical behavior and effects of compatibilization. *Comptes Rendus Mecanique*, **336**,

714-721.

29. Zhu, L., Fan, H.-N., Yang, Z.-Q., and Xu, X.-H. (2010) Evaluation of phase morphology, rheological, and mechanical properties based on polypropylene toughened with poly (ethylene-co-octene). *Polymer-Plastics Technology and Engineering*, **49**, 208-217.

30. Wang, J., Guo, J., Li, C., Yang, S., Wu, H., and Guo, S. (2014) Crystallization kinetics behavior, molecular interaction, and impact-induced morphological evolution of polypropylene/poly (ethylene-co-octene) blends: insight into toughening mechanism. *Journal of Polymer Research*, **21**, 618.

31. Liu, G., and Qiu, G. (2013) Study on the mechanical and morphological properties of toughened polypropylene blends for automobile bumpers. *Polymer Bulletin*, **70**, 849-857.

32. Yamaguchi, M., and Nitta, K. H. (1999) Optical and acoustical investigation for plastic deformation of isotactic polypropylene/ethylene-1-hexene copolymer blends. *Polymer Engineering and Science*, **39**, 833-840.

33. Ramírez-Vargas, E., Navarro-Rodríguez, D., Huerta-Martínez, B., Medellín-rodríguez, F., and Lin, J. (2000) Morphological and mechanical properties of polypropylene [PP]/poly (ethylene vinyl acetate)[EVA] blends. I. Homopolymer PP/EVA systems. *Polymer Engineering and Science*, **40**, 2241-2250.

34. Larbi, S., Bensaada, R., Bilek, A., and Djebali, S. (2015) Hygrothermal ageing effect on mechanical properties of FRP laminates. *AIP Conference Proceedings*, **1653**, 020066.

35. Han, M.-H., and Nairn, J. A. (2003) Hygrothermal aging of polyimide matrix composite laminates. *Composites, Part A: Applied Science and Manufacturing*, **34**, 979-986.

36. Hu, Y., Li, X., Lang, A. W., Zhang, Y., and Nutt, S. R. (2016) Water immersion aging of polydicyclopentadiene resin and glass fiber composites. *Polymer Degradation and Stability*, **124**, 35-42.

37. Retegi, A., Arbelaiz, A., Alvarez, P., Llano-Ponte, R., Labidi, J., and Mondragon, I. (2006) Effects of hygrothermal ageing on mechanical properties of flax pulps and their polypropylene matrix composites. *Journal of Applied Polymer Science*, **102**, 3438-3445.

38. Guermazi, N., Tarjem, A. B., Ksouri, I., and Ayedi, H. F. (2016) On the durability of FRP composites for aircraft structures in hygrothermal conditioning. *Composites, Part B: Engineering*, **85**, 294-304.

39. Le Gac, P.-Y., Arhant, M., Le Gall, M., and Davies, P. (2017) Yield stress changes induced by water in polyamide 6: characterization and modeling. *Polymer Degradation and Stability*, **137**, 272-280.

40. Mourad, A.-H., Fouad, H., and Elleithy, R. (2009) Impact of some environmental conditions on the tensile, creep-recovery, relaxation, melting and crystallinity behaviour of UHMWPE-GUR 410-

medical grade. *Materials and Design*, **30**, 4112-4119.

41. He, W., Liu, N., Chen, X., Guo, J., and Wei, T. (2016) The influence of hygrothermal ageing on the mechanical properties and thermal degradation kinetics of long glass fibre reinforced polyamide 6 composites filled with sepiolite. *RSC Advances*, **6**, 36689-36697.

42. Chen, Y., Davalos, J. F., Ray, I., and Kim, H.-Y. (2007) Accelerated aging tests for evaluations of durability performance of FRP reinforcing bars for concrete structures. *Composite Structures*, **78**, 101-111.

43. Ishak, Z. M., Ishiaku, U., and Karger-Kocsis, J. (2000) Hygrothermal aging and fracture behavior of short-glass-fiber-reinforced rubber-toughened poly (butylene terephthalate) composites. *Composites Science and Technology*, **60**, 803-815.

44. Islam, M. S., Pickering, K. L., and Foreman, N. J. (2010) Influence of hygrothermal ageing on the physico-mechanical properties of alkali treated industrial hemp fibre reinforced polylactic acid composites. *Journal of Polymers and the Environment*, **18**, 696-704.

45. Arhant, M., Le Gac, P.-Y., Le Gall, M., Burtin, C., Briançon, C., and Davies, P. (2016) Effect of sea water and humidity on the tensile and compressive properties of carbon-polyamide 6 laminates. *Composites, Part A: Applied Science and Manufacturing*, **91**, 250-261.

46. Shi, S., Chen, G., Wang, Z., and Chen, X. (2013) Mechanical properties of Nafion 212 proton exchange membrane subjected to hygrothermal aging. *Journal of Power Sources*, **238**, 318-323.

47. Le Gac, P.-Y., Choqueuse, D., Paris, M., Recher, G., Zimmer, C., and Melot, D. (2013) Durability of polydicyclopentadiene under high temperature, high pressure and seawater (offshore oil production conditions). *Polymer Degradation and Stability*, **98**, 809-817.

48. Wan, Y., Wang, Y., Huang, Y., He, B., and Han, K. (2005) Hygrothermal aging behaviour of VARTMed three-dimensional braided carbon-epoxy composites under external stresses. *Composites, Part A: Applied Science and Manufacturing*, **36**, 1102-1109.

49. Silva, L., Tognana, S., and Salgueiro, W. (2013) Study of the water absorption and its influence on the Young's modulus in a commercial polyamide. *Polymer Testing*, **32**, 158-164.

50. Guadagno, L., Fontanella, C., Vittoria, V., and Longo, P. (1999) Physical aging of syndiotactic polypropylene. *Journal of Polymer Science, Part B: Polymer Physics*, **37**, 173-180.

51. Guermazi, N., Elleuch, K., Ayedi, H., and Kapsa, P. (2008) Aging effect on thermal, mechanical and tribological behaviour of polymeric coatings used for pipeline application. *Journal of Materials Processing Technology*, **203**, 404-410.

52. Berthé, V., Ferry, L., Bénézet, J., and Bergeret, A. (2010) Ageing of different biodegradable polyesters blends mechanical and hygrothermal behavior. *Polymer Degradation and Stability*, **95**, 262-269.

53. Gautier, L., Mortaigne, B., Bellenger, V., and Verdu, J. (2000) Osmotic cracking nucleation in hydrothermal-aged polyester matrix. *Polymer*, **41**, 2481-2490.

54. Boubakri, A., Haddar, N., Elleuch, K., and Bienvenu, Y. (2010) Impact of aging conditions on mechanical properties of thermoplastic polyurethane. *Materials and Design*, **31**, 4194-4201.

55. Boubakri, A., Elleuch, K., Guermazi, N., and Ayedi, H. (2009) Investigations on hygrothermal aging of thermoplastic polyurethane material. *Materials and Design*, **30**, 3958-3965.

56. Mondal, S., Hu, J., and Yong, Z. (2006) Free volume and water vapor permeability of dense segmented polyurethane membrane. *Journal of Membrane Science*, **280**, 427-432.

57. Deshmane, C., Yuan, Q., Perkins, R., and Misra, R. (2007) On striking variation in impact toughness of polyethylene–clay and polypropylene–clay nanocomposite systems: the effect of clay–polymer interaction. *Materials Science and Engineering: A*, **458**, 150-157.

58. Guermazi, N., Elleuch, K., and Ayedi, H. (2009) The effect of time and aging temperature on structural and mechanical properties of pipeline coating. *Materials and Design*, **30**, 2006-2010.

59. *Handbook of Plastics Analysis*, Lobo, H., and Bonilla, J. V. (eds.), Taylor & Francis, USA (2003).

60. Zhao, S., Chen, F., Zhao, C., Huang, Y., Dong, J.-Y., and Han, C. C. (2013) Interpenetrating network formation in isotactic polypropylene/graphene composites. *Polymer*, **54**, 3680-3690.

61. Li, C.-Q., Zha, J.-W., Long, H.-Q., Wang, S.-J., Zhang, D.-L., and Dang, Z.-M. (2017) Mechanical and dielectric properties of graphene incorporated polypropylene nanocomposites using polypropylene-graft-maleic anhydride as a compatibilizer. *Composites Science and Technology*, **153**, 111-118.

62. Gopakumar, T., and Pagé, D. (2004) Polypropylene/graphite nanocomposites by thermo-kinetic mixing. *Polymer Engineering and Science*, **44**, 1162-1169.

63. Panda, B. P., Mohanty, S., and Nayak, S. K. (2015) Mechanism of toughening in rubber toughened polyolefin - A review. *Polymer-Plastics Technology and Engineering*, **54**, 462-473.

64. Bu, H., Qiu, W., Tan, Z., Li, Q., and Rong, J. (2015) Study on toughening of poly (4-methyl-1-pentene) with various thermoplastic elastomers. *Journal of Thermoplastic Composite Materials*, **28**, 1334-1342.

65. Chow, C., Xing, X., and Li, R. (2007) Moisture absorption studies of sisal fibre reinforced polypropylene composites. *Composites Science and Technology*, **67**, 306-313.

66. Zanni-Deffarges, M., and Shanahan, M. (1995) Diffusion of water into an epoxy adhesive: comparison between bulk behaviour and adhesive joints. International *Journal of Adhesion and Adhesives*,

15, 137-142.

67. Chu, W., Wu, L., and Karbhari, V. M. (2004) Durability evaluation of moderate temperature cured E-glass/vinyl ester systems. *Composite Structures*, **66**, 367-376.

68. Parameswaranpillai, J., Joseph, G., Shinu, K., Sreejesh, P., Jose, S., Salim, N. V., and Hameed, N. (2015) The role of SEBS in tailoring the interface between the polymer matrix and exfoliated graphene nanoplatelets in hybrid composites. *Materials Chemistry and Physics*, **163**, 182-189.

69. Ferrer, G. G., Sánchez, M. S., Sánchez, E. V., Colomer, F. R., and Ribelles, J. L. G. (2000) Blends of styrene–butadiene–styrene triblock copolymer and isotactic polypropylene: morphology and thermomechanical properties. *Polymer International*, **49**, 853-859.

70. Liao, C. Z., and Tjong, S. C. (2011) Effects of carbon nanofibers on the fracture, mechanical, and thermal properties of PP/SEBS-g-MA blends. *Polymer Engineering and Science*, **51**, 948-958.

7

Hygrothermal Ageing of Isotactic Polypropylene and Impact Polypropylene Copolymers: A Comparison of Mechanical, Thermal, Structural and Morphological Properties

7.1 Introduction

Thermoplastics are a promising class of polymers which can be re-processed and reused by thermal means for a myriad of applications spanning over both engineering and non-engineering sectors owing to their tailorable properties, ease of processing, light weight, low cost, corrosion resistance, etc. [1-4]. Specifically, polypropylene (PP) is a versatile polyolefin thermoplastic material available in various forms like homopolymer, copolymer, reactor-blends, etc. As com-pared to other thermoplastics, the properties of PP can be relatively easily tuned through simple modifications in its structural arrange-ment [5-8].

Isotactic polypropylene (iPP), a homopolymer PP with same side methyl group alignment, is the most common PP grade accounting for more than 20 vol% of total PP production. In addition to usual benefits of PP such as low price and density as well as high re-sistance against corrosion and thermal degradation, iPP exhibits good balance of mechanical and physical characteristics along with superior recyclability due to the combinatorial effect of its isotactic nature and narrow distribution of molecular weight [9,10]. Howev-er, iPP displays poor impact performance on subjecting to sudden mechanical force due to its low crack growth resistance, especially at sub-ambient temperatures [11,12]. The introduction of various macromolecular materials such as elastomers, thermoplastic elas-tomers, copolymers, etc., to iPP has been reported to result in en-hanced impact performance, thereby, paving the way for use in en-gineering applications [13-25]. Recently, we have also reported simultaneously improved impact strength and tensile modulus for

Anish Varghese and Vikas Mittal*, The Petroleum Institute (part of Khalifa University of Science and Technology), Abu Dhabi, UAE
*Current address: Bletchington, Wellington County, Australia

iPP through the use of a combined system of maleic anhydride-*graft*-poly(ethylene-vinyl acetate) and graphene [23].

Impact polypropylene copolymers (IPC), also named as heterophasic copolymers, have been developed to overcome the poor impact performance associated with iPP. For this, the reactor induced copolymerization of iPP is carried out with ethylene monomer *in-situ*. Borstar and Spheripol are the two widely accepted processes for this purpose, which comprise iPP bulk polymerization and subsequent propylene-ethylene copolymerization in the gas phase [26-30]. IPC comprise of a multiphase system with iPP crystalline homopolymer, ethylene-propylene semi-crystalline block copolymer (EPBC) and ethylene-propylene amorphous random copolymer (EPRC) [31-33]. Among these, EPRC can act as effective stress concentrator on the iPP surface, whereas EPBC is responsible for strengthening the interfacial performance between the other two by serving as a robust compatibilizer. On exposing IPC to the impact force, the stress concentrator phase, EPRC, consumes a majority of the force, leading to the formation of micro-voids on the iPP surface by means of crazing and shear yielding, thus, resulting in plastic deformation and resistance to crack propagation. The ultimate impact performance of IPC greatly depends on the size, shape and distribution of EPRC, along with its miscibility with iPP phase under the influence of EPBC [32,34-36].

During their service life, the materials are occasionally subjected to harsh environmental conditions such as temperature and pressure variations, humidity, sunlight, marine environment, chemicals and pollutants, mechanical impact, etc. [37-39]. In order to estimate the endurance of the materials under harsh environmental conditions, laboratory accelerated ageing environments imitating the real-life harsh conditions can be employed, thus, avoiding long-term testing [40,41]. Among different types of ageing, hygrothermal ageing causes structural changes in the polymers, which negatively affect the service life owing to the reduction in mechanical and thermal characteristics [42-44]. The extent of hygrothermal ageing induced changes in the polymers is closely related to the ageing conditions such as medium, temperature and time [45-51].

In this study, the behavior of three PP grades was studied under the influence of hygrothermal environment. The test specimens were prepared by injection molding and were subsequently immersed in both distilled water (DW) and saline water (SW) for 32 days at 70 and 90 °C. In order to assess the water uptake with ageing time, the sample weight was measured periodically after 4 days. Overall, hygrothermal ageing induced changes in the mechanical, thermal, structural and morphological properties of the polymer grades were analyzed.

7.2 Experimental

7.2.1 Materials

Isotactic polypropylene grade and two impact polypropylene copolymer grades (named as IPC1 and IPC2) were provided by Abu Dhabi Polymers Company Limited, UAE (in pellet form) and were used as received without any subsequent treatment. Table 7.1 presents further details of these materials.

Table 7.1 Material data for iPP, IPC1 and IPC2

Material	Nature/ Constituents of the Polymers	Appearance	Details
Isotactic polypropylene (iPP)	Homopolymer/iPP		**Density:** 900-910 kg.m^{-3} **Melt flow rate:** 8 g/10 min (230 °C, 2.16 kg) **Melting point:** 168.5 °C
Impact polypropylene copolymer 1 (IPC1)	Heterophasic copolymer/iPP, EPRC and EPBC		**Density:** 900-910 kg.m^{-3} **Melt flow rate:** 38 g/10 min (230 °C, 2.16 kg) **Melting point:** 166.5 °C
Impact polypropylene copolymer 2 (IPC2)	Same as for IPC1		**Density:** 900-910 kg.m^{-3} **Melt flow rate:** 7 g/10 min (190 °C, 2.16 kg) **Melting point:** 167.7 °C

7.2.2 Preparation of Samples

In order to proceed with hygrothermal ageing, the PP samples were processed into dumbbell and rectangular bar shaped tensile and impact test specimens. For this purpose, a lab-scale plunger type injection molding machine HAAKE MiniJet PRO (Thermo Scientific, Germany) was employed. The test specimens were prepared by using the conditions: 180 °C cylinder temperature with a holding time of 5 min to obtain uniform PP melt, 430 bar injection pressure for 10 s to uniformly fill the mold volume and 500 bar holding pressure for 6 s to prevent backflow. The mold was maintained at a temperature of 125 °C.

7.2.3 Hygrothermal Ageing of Samples

To simulate the harsh environmental conditions, accelerated hygrothermal ageing in distilled water as well as synthetic saline water was carried out, in accordance with ASTM D570-90. Specifically, the accelerated ageing in DW and SW simulated humid and saline environments respectively. For this purpose, the samples were immersed in the media for 32 days at 70 and 90 °C. The concentration of NaCl in SW was fixed at 3%.

7.2.4 Characterization of Unaged and Aged Samples

Water Uptake

Water uptake, also termed as weight gain ratio (W_g), of the hygrothermally immersed PP samples was estimated using the unaged sample weight (W_0) and the weight of the aged sample (W_t) measured with a gap of 4 days, as per the following equation:

$$W_g = \frac{W_t - W_0}{W_0} \times 100$$

For measuring W_t, the samples were taken out from the immersed medium and weighed after wiping away the surface moisture using a neat cotton towel. The reported W_0 and W_t values constitute the average of at least five samples with less than 10% standard deviation.

Fourier-transform Infrared (FTIR) Spectroscopy

Ageing induced structural changes in the PP samples were analyzed using Fourier-transform infrared spectra, generated in transmission mode by employing BRUKER TENSOR II series spectrometer. With the aid of a diamond attenuated total reflectance (ATR) crystal, thin surface portion of the injection molded PP samples were examined by collecting 32 scans. The wavenumber range of 4000-400 cm^{-1} and a resolution of 4 cm^{-1} were used.

Wide-angle X-ray Diffraction (WAXD)

X'Pert PRO Panalytical powder diffractometer was used to study the diffraction patterns of the injection molded PP samples (at 40 mA and 45 kV). 2-theta angle range of 5-60° was scanned with a speed of 0.017° s^{-1} and a step time of 10 s. The analysis was carried out under ambient conditions using Cu-Kα irradiation ($\lambda = 1.5406$ Å).

Tensile Properties

Instron 3345 universal testing machine (UTM), attached with a load cell of 50 kN, was employed to study the impact of hygrothermal ageing on the tensile performance of the PP samples under ambient conditions (crosshead speed of 10 mm.min^{-1}). The dumbbell-shaped test specimens with dimensions 75 mm x 5 mm x 2 mm and a span length of 35 mm were used according to ISO 527 standard, and the reported data constitutes the mean of values from five specimens.

Izod Impact Strength

Resil impactor (Ceast) was employed to study the Izod impact strength of the unaged and aged samples at room temperature. The bar shaped test specimens with dimensions 80 mm x 10 mm x 4 mm were used according to ISO 180 standard. The hammer speed used for the analysis was 3.64 m.s^{-1}, and the reported data represents the average of five specimens.

Differential Scanning Calorimetry

Differential scanning calorimeter from TA Instruments was used to study the peak melting temperature (T_m), peak crystallization temperature and (T_c), melting enthalpy (ΔH_f), crystallization enthalpy (ΔH_c) and percentage crystallinity (X_c). 3-8 mg sample weight was examined in the temperature range of (-50, +200 °C) by applying two heating and cooling cycles using a dry nitrogen flow of 50 mL.min[-1]. Second heating-cooling cycles were used to record the calorimetric behavior of the samples.

Scanning Electron and Optical Microscopy

Scanning electron microscope (SEM) FEI Quanta (FEG250, USA) was used to analyze the micro-morphology of the impact fractured surface of the PP samples. The images were taken at 10 kV after sputtering a thin layer of gold on the fracture surfaces. The samples were affixed on the aluminum stubs for analysis using conductive carbon tape. Optical microscope OLYMPUS BX51M (Japan) was also used to examine the surface changes induced by hygrothermal conditions (eyepiece magnification of 10x).

7.3 Results and Discussion

Figure 7.1 depicts the change in the extent of water uptake as a function of hygrothermal ageing time for iPP, IPC1 and IPC2, at 70 and

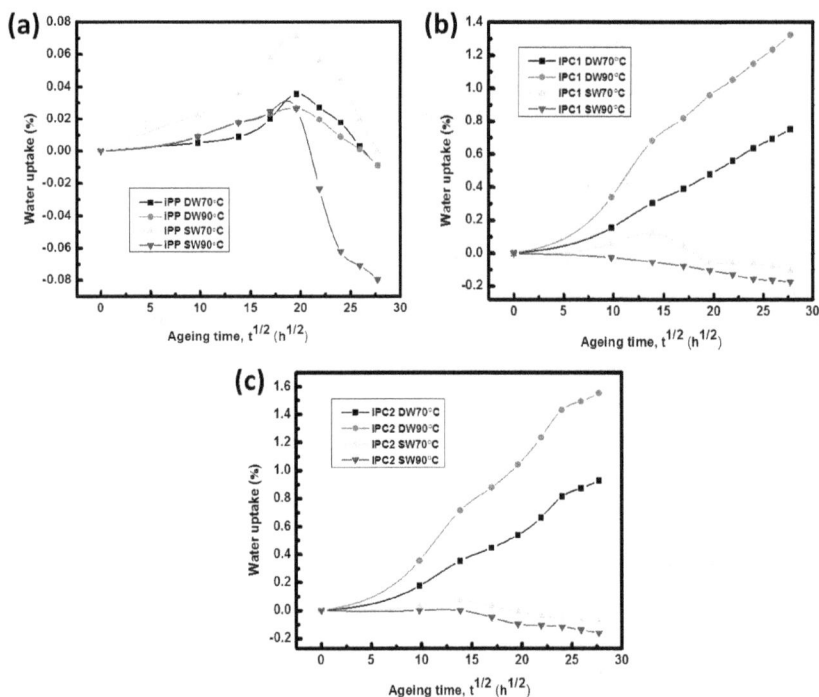

Figure 7.1 Water uptake curves of (a) iPP, (b) IPC1 and (c) IPC2.

90 °C in DW and SW. The water uptake behavior was observed to be different for iPP and IPC. Water uptake curves of iPP (Figure 7.1a) displayed two distinct parts, which supported the presence of combinatorial Fickian and non-Fickian behaviors regardless of the ageing conditions. The Fickian behavior, represented by an increasing trend in water uptake with time in a non-linear manner upon hygrothermal ageing, was visible until ageing time ($\sqrt{h}$) = 19.6. Following this, a decreasing trend of water uptake with time represented the non-Fickian behavior. Comparing different hygrothermal conditions, iPP aged at 70 °C in SW displayed the highest water uptake, i.e., ca. 0.071%. The Fickian behavior is associated with the moisture diffusion in the samples under hygrothermal conditions, whereas the non-Fickian behavior might result from surface crosslinking, PP chain rearrangement and washing effect [52,53]. Both IPC samples (Figures 7.1b and 7.1c) demonstrated similar behavior upon hygrothermal immersion, though the observed behavior differed in DW and SW. In DW, the water uptake increased with ageing time non-linearly throughout the studied ageing time range regardless of the IPC type owing to the osmotic cracking effect induced by the hydrophilic oligomer moieties [54,55]. Also, such effect was observed to increase on enhancing the temperature from 70 to 90 °C due to the rapid reorganization of the loose macromolecules and, thus, the appearance of increased free volume [56,57]. A higher water uptake value of 1.55% was observed for IPC2 in DW at 90 °C as compared to 1.32% for IPC1. In SW, the observed water uptake values at 70 °C were extremely low for both IPC1 and IPC2 samples in comparison with the samples aged in DW. On the other hand, at 90 °C, the observed water uptake values were negative. The appearance of such behavior in SW is related to the reduction in the chemical potential contributed by salt [53]. In SW at 70 °C, the water uptake values increased only until $\sqrt{h}$ = 13.85 for both IPC1 and IPC2, followed by a decreasing trend which even reached negative values after $\sqrt{h}$ = 19.59. Comparing the two IPC samples, IPC2 demonstrated lower water uptake in SW. From these observations, it can be seen that iPP offered superior barrier performance in DW, whereas IPC1 and IPC2 were better in SW.

FTIR spectra were collected for samples before and after simulated hygrothermal ageing in DW and SW to assess the structural changes, as shown in Figure 7.2. For unaged iPP, characteristic peaks at 2957, 2916, 2870, 2837, 2362, 1463, 1374, 999-973-844-811 and 898 cm^{-1} could be attributed to CH$_3$ asymmetric stretching

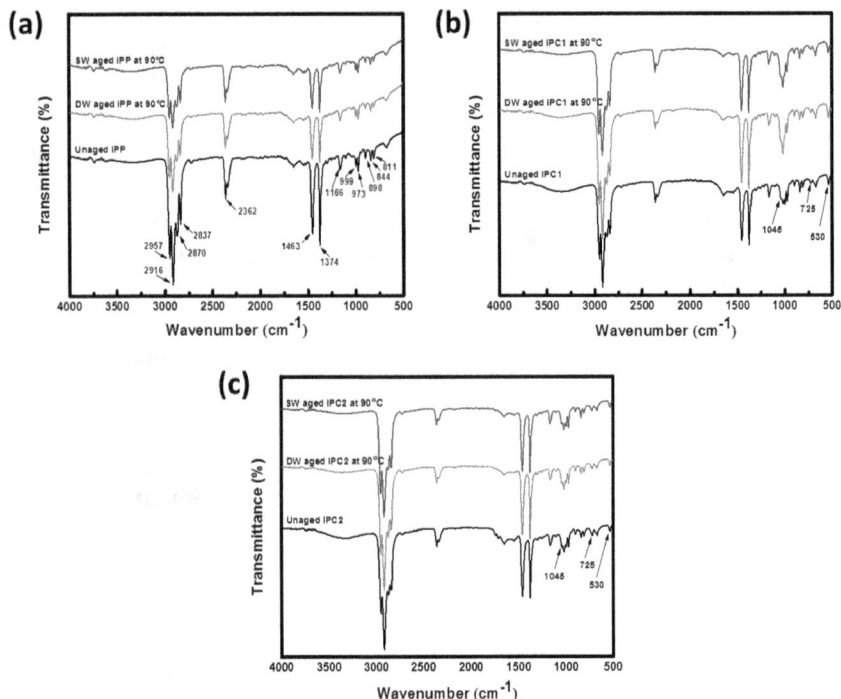

Figure 7.2 FTIR spectra for unaged and aged samples: (a) iPP, (b) IPC1 and (c) IPC2.

vibration, CH$_2$ symmetric stretching vibration, CH$_3$ symmetric stretching vibration, CH$_2$ symmetric stretching vibration, C-H bending vibration, CH$_3$ asymmetric bending vibration, CH$_3$ symmetric bending vibration, tertiary methyl skeleton deformation and stretching caused perpendicular absorption respectively [58]. FTIR spectra of 90 °C DW and SW aged iPP samples also displayed same characteristic peaks as unaged iPP, though the intensity of each peak was observed to reduce. Such intensity reduction indicates the initiation of chain scission under hygrothermal conditions. In the FTIR spectra of IPC1 (Figure 7.2b) and IPC2 (Figure 7.2c), three additional peaks at 1045, 725 and 530 cm^{-1} were observed along with the aforesaid characteristic iPP peaks. The additional peaks were contributed by the amorphous polyethylene composition in the IPC1 and IPC2 samples, which were attributed to CH$_2$ gauche configuration, CH$_2$ rocking vibration and CH$_2$ stretching vibration respectively [28,59-61]. After hygrothermal ageing at 90 °C, the characteristic peaks of IPC1 and IPC2 were observed to appear with increased intensity in both DW and SW. Specifically, DW aged IPC samples exhibited higher increase in the peak intensities as compared to SW. According to previous studies, thermal oxidation is the cause of such increase in the peak intensities after hygrothermal exposure [62,63]. Correspondingly, in the current study, a higher thermal oxidation tendency was confirmed in DW than SW for both IPC1 and IPC2 samples. Also, the observed higher extent of peak intensity increase in IPC1 underlines its greater tendency for thermal oxidation than IPC2.

Wide angle X-ray diffractograms of unaged and aged samples are depicted in Figure 7.3. For unaged iPP sample, the diffraction peaks

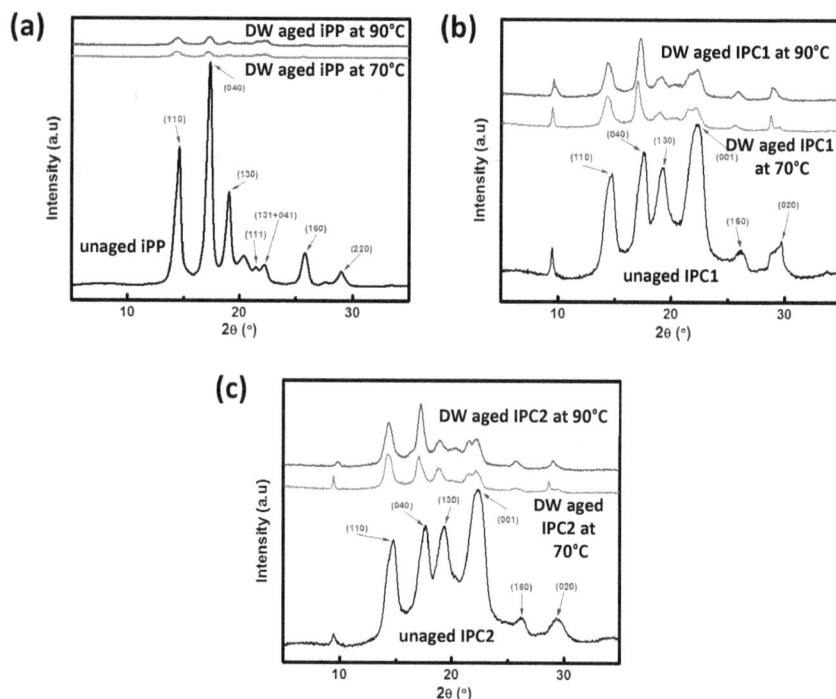

Figure 7.3 WAXD patterns for unaged and aged samples: (a) iPP, (b) IPC1 and (c) IPC2.

at 14.55°, 17.26°, 19.03°, 21.47°, 22.18°, 25.76° and 29.01° could be attributed to respective (110), (040), (130), (111), (131+041), (160) and (220) α-crystalline planes [23,64,65]. To estimate the interplanar distance (*d*) of each diffraction peak, Bragg's equation (*d* = $\lambda/(2\sin\theta_{max})$ where λ is the *X*-ray wavelength (1.5406 Å), was used.

Correspondingly, d values of 6.08, 5.13, 4.66, 4.14, 4.00, 3.46 and 3.08 Å respectively were obtained for the aforesaid α-crystalline planes. The unaged IPC1 and IPC2 exhibited α-crystalline peaks of iPP and PE attributed to respective (110), (040), (130), (160), (020) and (001) planes. These peaks were observed at 14.67°, 17.60°, 19.19°, 26.11°, 29.69° and 22.33° respectively for IPC1 and 14.77°, 17.72°, 19.29°, 26.22°, 29.36° and 22.33° respectively for IPC2. Accordingly, the obtained d values of these peaks were 6.03, 5.04, 4.62, 3.41, 3.01 and 3.98 Å respectively for IPC1 and 5.99, 5.00, 4.60, 3.39, 3.04 and 3.98 Å respectively for IPC2. Similar characteristic diffraction peaks were observed in the hygrothermally aged samples, however, with reduced intensity. The observed crystallinity reduction indicated the ageing induced structural changes in the samples owing to chain scission. However, the decrease in the diffraction peak intensity was observed to be lower at 90 °C as compared at 70 °C for the aged PP samples, regardless of their type, due to the rearrangement of the macromolecular chain structure at 90 °C [48].

The room temperature Izod impact strength of iPP, IPC1 and IPC2 was estimated before and after simulated hygrothermal ageing using un-notched specimens, as presented in Table 7.2. For unaged iPP, IPC1 and IPC2, impact strength of ca. 7, 53 and 103 kJ.m^{-2} respectively was observed. The impact strength of iPP was observed

Table 7.2 Mechanical properties of unaged and aged samples

Sample	Impact strength (kJ.m^2)	Tensile strength (MPa)	Tensile modulus (MPa)	Breaking extension (%)
Unaged iPP	7	39	1011	24
DW aged iPP at 70 °C	51	36	815	23
DW aged iPP at 90 °C	45	35	808	21
SW aged iPP at 70 °C	45	36	809	21
SW aged iPP at 90 °C	46	35	801	21
Unaged IPC1	53	22	1018	35
DW aged IPC1 at 70 °C	68	21	551	14
DW aged IPC1 at 90 °C	92	20	471	15
SW aged IPC1 at 70 °C	75	21	486	18
SW aged IPC1 at 90 °C	89	21	481	21
Unaged IPC2	103	21	864	299
DW aged IPC2 at 70 °C	101	19	270	120
DW aged IPC2 at 90 °C	101	19	205	163
SW aged IPC2 at 70 °C	102	18	259	113
SW aged IPC2 at 90 °C	102	18	220	154

to markedly enhance after ageing under hygrothermal conditions owing to the plasticization effect [43,66]. Accordingly, the observed impact strength retention rate of iPP stayed above 550% (Figure 7.4a) regardless of the ageing conditions. The impact strength of the

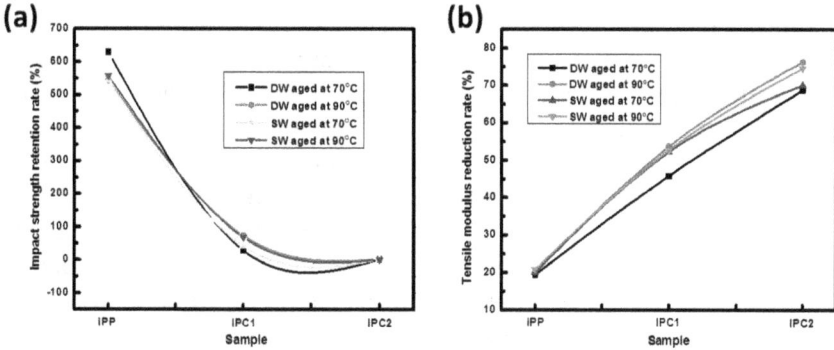

Figure 7.4 Effect of hygrothermal conditions on (a) impact strength retention rate and (b) tensile modulus reduction rate.

aged IPC1 also exhibited an increasing trend, which was also observed to increase with increase in temperature from 70 to 90°C in both DW and SW. For instance, the impact strength retention rates of ca. 74 and 68% were discerned for IPC1 after hygrothermal ageing at 90 °C in DW and SW respectively. The observed increase in the impact strength retention at higher temperature is related to the swift micro-voids formation at the iPP-EPRC interface in the IPC1 sample owing to the increasing degree of water plasticization [43,48]. The impact strength of IPC2 was observed to be retained after hygrothermal ageing. Thus, the impact strength values of 101 and 102 kJ.m^{-2} were discerned for IPC2 after hygrothermal ageing at 90 °C in DW and SW respectively, as compared to 103 kJ.m^{-2} for the unaged IPC2. Slight reduction in the impact strength indicates the beginning of the structural integrity loss under the influence harsh hygrothermal conditions [48].

The tensile modulus values of 1011, 1018, and 864 MPa were measured for unaged iPP, IPC1 and IPC2 respectively. After hygrothermal ageing, the tensile modulus of the PP samples was observed to decrease. In addition, the tensile modulus reduction rate increased with temperature in both DW and SW. For iPP, the observed tensile modulus reduction rate was higher in SW at both 70 and 90 °C as compared to DW due to the influence of high pH water medium and increased crack formation tendency [38,67]. An optimum reduction rate of ca. 21% (Figure 7.4b) was recorded for 90 °C SW aged iPP. For aged IPC1 and IPC2, higher reduction rates were observed at 70 °C in SW and at 90 °C in DW. Under hygrothermal ageing conditions, the reduction in the tensile modulus probably results from the macromolecular chain scission and the resultant decrease in the material resistance. In addition, owing to the increase in water penetration with temperature, the aforesaid effect would become more prominent [68,69]. After hygrothermal ageing, a reduction in the tensile strength of the PP samples was also observed. Specifically, the tensile strength reduction rates for iPP, IPC1 and IPC2 were observed to be 10, 9 and 10% respectively in DW at 90 °C and 10, 5 and 14% respectively in SW at 90 °C. Similarly, the extension at break of the samples was also negatively affected after hygrothermal exposure. The extension at break of iPP was observed to reduce with temperature in DW, though the reduction rate remained unchanged in SW at both 70 and 90 °C (Table 7.2). For IPC1 and IPC2, hygrothermal ageing had more negative effect on the extension at 70 °C in both DW and SW than 90 °C. Overall, the maximal extension reduction rates of 13, 60 and 62% were noticed for iPP, IPC1 and IPC2 respectively. From these observations, lower resistance of IPC samples against extension reduction in hygrothermal conditions as compared to iPP was confirmed due to the influence of amorphous content and, thus, higher water penetration tendency and molecular scission [70].

The calorimetric properties of the samples are depicted in Table 7.3. The second heating and cooling curves of the samples are also demonstrated in Figures 7.5 and 7.6. The peak melting temperature of iPP, IPC1 and IPC2 was observed to be 168.5, 166.5 and 167.7 °C respectively. For aged iPP, the observed T_m was slightly lower in DW and almost same in SW, as compared to the unaged polymer. In the case of aged IPC1, T_m was slightly higher in all hygrothermal conditions as compared to the unaged sample. For IPC2, a slight decrease in T_m was observed with temperature in both DW and SW. These observations indicated that the PP samples resisted severe changes in T_m upon hygrothermal exposure. The peak crystallization temperatures of 127.9, 130.4 and 128.9 °C were measured for unaged iPP, IPC1 and IPC2. For aged IPC1, T_c was slightly increased in DW with temperature, whereas it remained unchanged in SW, though still higher than unaged IPC1. For aged IPC2, T_c displayed a slight reduction of ≤0.5 °C in both DW and SW.

Table 7.3 Calorimetric properties of unaged and aged iPP, IPC1 and IPC2

Sample	T_c (°C)	ΔH_c (J/g)	T_m (°C)	ΔH_f (J/g)	X_c (%)	ΔX_c (%)
Unaged iPP	127.9	98.7	168.5	90.8	43.8	-
DW aged iPP at 70 °C	126.3	106.2	166.9	112.2	54.2	10.4
DW aged iPP at 90 °C	126.4	110.1	166.9	108.0	52.1	8.3
SW aged iPP at 70 °C	126.2	111.2	168.2	106.5	51.4	7.6
SW aged iPP at 90 °C	126.2	107.0	168.6	105.4	50.9	7.1
Unaged IPC1	130.4	84.3	166.5	82.8	40.0	-
DW aged IPC1 at 70 °C	130.9	88.9	166.9	93.8	45.3	6.3
DW aged IPC1 at 90 °C	131.2	92.0	166.6	91.6	44.2	5.2
SW aged IPC1 at 70 °C	130.9	91.7	166.7	91.1	44.0	4.0
SW aged IPC1 at 90 °C	130.9	90.3	166.9	90.8	43.8	3.8
Unaged IPC2	128.9	74.5	167.7	72.9	35.2	-
DW aged IPC2 at 70 °C	128.5	78.2	167.3	79.0	38.2	3.0
DW aged IPC2 at 90 °C	128.6	79.8	166.7	78.5	37.9	2.7
SW aged IPC2 at 70 °C	128.4	79.5	167.5	77.8	37.5	2.3
SW aged IPC2 at 90 °C	128.4	79.5	167.1	76.6	37.0	1.8

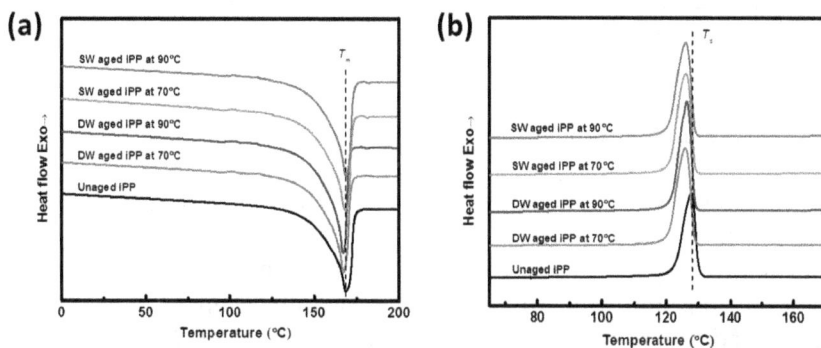

Figure 7.5 DSC (a) heating and (b) cooling thermograms of unaged and hygrothermally aged iPP.

For estimating X_c, the following equation was used [24,55]:

$$X_c = \frac{\Delta H}{\Delta H_0} \times 100$$

In this equation, ΔH_0, and ΔH correspond to theoretically calculated ΔH_f (207.1 J/g) for totally crystalline PP [24] and ΔH_f for the analyzed samples from DSC.

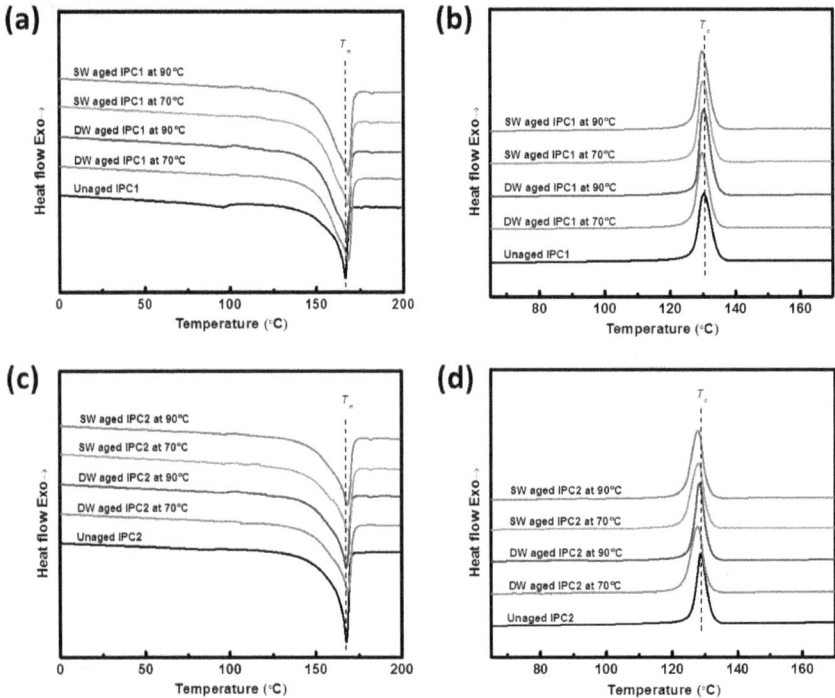

Figure 7.6 DSC heating curves for unaged and aged (a) IPC1 and (c) IPC2; DSC cooling curves for unaged and aged (b) IPC1 and (d) IPC2.

As depicted in Table 7.3, X_c values of 43.8, 40.0 and 35.2 were calculated for unaged iPP, IPC1 and IPC2 respectively. After hygrothermal ageing, X_c values were observed to increase, regardless of PP type, possibly due to the chemi-crystallization of the initially scissioned macromolecules [55]. A higher extent of X_c increase was observed at 70 °C than 90 °C in both DW and SW. In addition, the samples exhibited higher rate of increase in DW as compared to SW. Specifically, the extent of X_c increase was aligned in the range of 7.1-10.4%, 2.8-6.3% and 1.8-3.0% for iPP, IPC1 and IPC2 respectively.

Figure 7.7 presents the SEM images for unaged and aged iPP, IPC1 and IPC2 samples. The SEM images of the unaged iPP, 70 °C DW aged iPP and 90 °C DW aged iPP (Figure 7.7a-c) supported the transition of the iPP fracture morphology from brittle to ductile upon hygrothermal exposure due to the effect of the water plasticizing effect. These observations also supported the earlier observed impact properties of the samples. The SEM images of the unaged IPC1 and IPC2 (Figures 7.7d and 7.7g) exhibited several round-shaped

micro-voids in the continuous phase. The observed micro-voids confirmed the effective debonding or cavitation at the EPRC-iPP

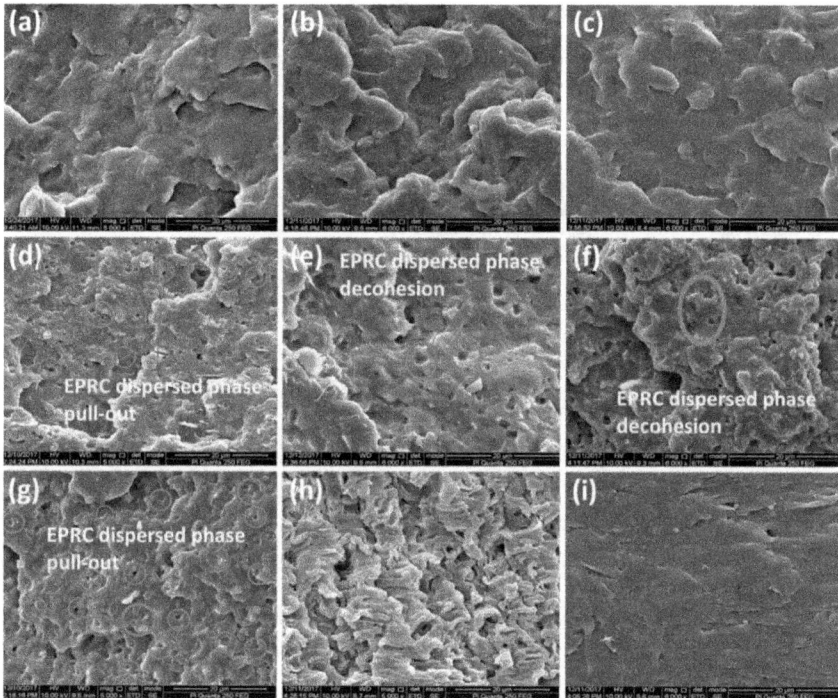

Figure 7.7 SEM images of the impact fractured surface for unaged (a) iPP,
(d) IPC1 and (g) IPC2; 70 °C DW aged (b) iPP (e) IPC1 and (h) IPC2; and 90
°C DW aged (c) iPP, (f) IPC1 and (i) IPC2.

interface owing to shear yielding and crazing upon impact. Accordingly, the increased impact strength values of 53 and 103 kJ.m^{-2}
were observed for the corresponding IPC1 and IPC2 samples. By
virtue of better EPRC-iPP interfacial performance under the influence of EPBC, IPC2 exhibited higher impact strength than IPC1. Figure 7.7e-f demonstrate the impact fracture morphology of DW aged
IPC1 at 70 and 90 °C respectively. Decohesion of the EPRC phase in
the continuous matrix was evident after hygrothermal exposure at
both 70 and 90 °C. As a result, slightly bigger round-shaped EPRC
phase pull-outs were observed to appear. Figure 7.7h-i show the
impact fracture morphology of the hygrothermally aged IPC2 at 70
and 90 °C respectively. The initiation of the phase damage of the
dispersed EPRC phase was observed after hygrothermal exposure.

Accordingly, slightly reduced impact strength of 101 kJ.m^{-2} was observed at both 70 and 90 °C for the aged IPC2 sample.

Optical micrographs of the hygrothermally aged samples at 70 and 90 °C in DW are depicted in Figure 7.8. Surface uniformity loss and resultant deterioration were observed for all samples. Accordingly, the occurrence of pore analogous structure with irregular shapes and minor surface cracks was observed, which underlined the increasing trend of surface roughness with hygrothermal ageing. Among the studied samples, IPC1 suffered more surface damage upon ageing as compared to IPC2 and iPP.

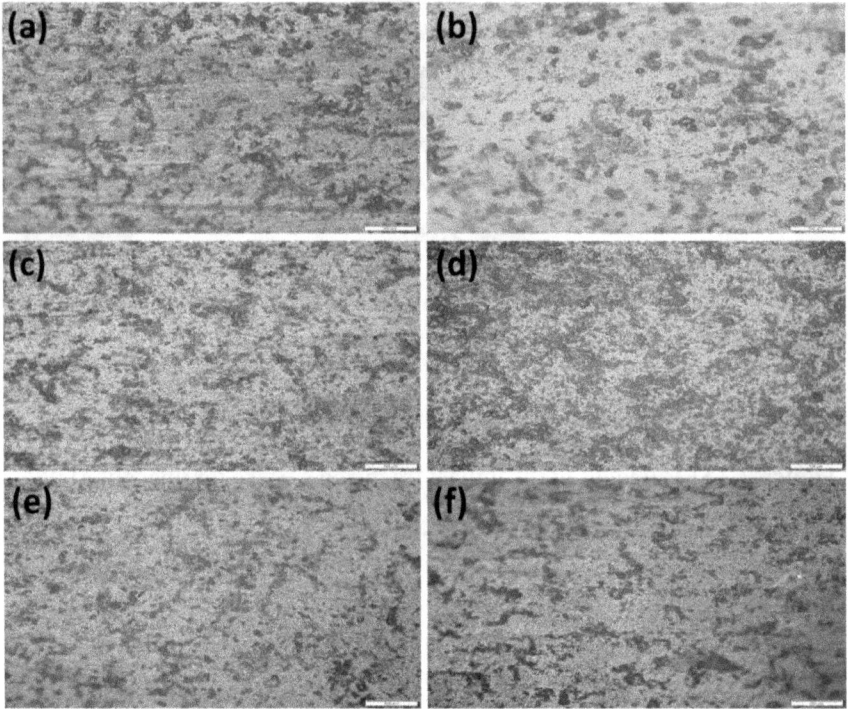

Figure 7.8 Optical micrographs of 70 °C DW aged (a) iPP, (c) IPC1 and (e) IPC2; and 90 °C DW aged (b) iPP, (d) IPC1 and (f) IPC2.

7.4 Conclusions

In this study, hygrothermal ageing analysis of isotactic polypropylene (iPP) and two different impact polypropylene copolymers (IPC1 and IPC2) was carried out using different ageing conditions. The iPP polymer displayed superior barrier performance in DW, though IPC samples were better in SW. Significantly enhanced values of impact strength were recorded after hygrothermal exposure for iPP and IPC1 due to the water plasticizing effect, whereas a slight reduction was noted for IPC2. The impact fracture morphology of the samples also supported the observations from the impact analysis. Owing to the combinatorial thermal and water ageing effect and the resultant molecular chain scission, the tensile modulus, strength and elongation at break of the aged samples were observed to decrease. As a result of the chemi-crystallization phenomenon, increased percentage crystallinity values were discerned for the PP samples.

References

1. Mortazavian, S., Fatemi, A., Mellott, S. R., and Khosrovaneh, A. (2015) Effect of cycling frequency and self-heating on fatigue behavior of reinforced and unreinforced thermoplastic polymers. *Polymer Engineering and Science*, **55**, 2355-2367.
2. Teoh, S., Tang, Z., and Hastings, G. W. (2016) Thermoplastic polymers in biomedical applications: Structures, properties and processing. In: *Handbook of Biomaterial Properties*, Black, J., and Hastings, G. (eds.), Springer, USA, pp. 261-290.
3. Paoletti, A., Lambiase, F., and Di Ilio, A. (2016) Analysis of forces and temperatures in friction spot stir welding of thermoplastic polymers. *The International Journal of Advanced Manufacturing Technology*, **83**, 1395-1407.
4. Eftekhari, M., and Fatemi, A. (2016) Tensile behavior of thermoplastic composites including temperature, moisture, and hygrothermal effects. *Polymer Testing*, **51**, 151-164.
5. Maddah, H. A. (2016) Polypropylene as a promising plastic: A review. *American Journal of Polymer Science*, **6**, 1-11.
6. Gahleitner, M., and Paulik, C. (2017) Polypropylene and other polyolefins. In: *Brydson's Plastics Materials*, 8th edition, Elsevier, USA, pp. 279-309.
7. Tripathi, D. (2002) *Practical Guide to Polypropylene*, iSmithers Rapra Publishing, USA.
8. *Handbook of Polypropylene and Polypropylene Composites*, Karian, H. G. (ed.), CRC Press, USA (2003).
9. Bikiaris, D., Vassiliou, A., Chrissafis, K., Paraskevopoulos, K., Jannakoudakis, A., and Docoslis, A. (2008) Effect of acid treated multi-walled carbon nanotubes on the mechanical, permeability, thermal properties and thermo-oxidative stability of isotactic polypropylene. *Polymer Degradation and Stability*, **93**, 952-967.

10. Wang, Y., and Tsai, H. B. (2012) Thermal, dynamic-mechanical, and dielectric properties of surfactant intercalated graphite oxide filled maleated polypropylene nanocomposites. *Journal of Applied Polymer Science*, **123**, 3154-3163.

11. Chen, F., Qiu, B., Wang, B., Shangguan, Y., and Zheng, Q. (2015) Balanced toughening and strengthening of ethylene–propylene rubber toughened isotactic polypropylene using a poly (styrene-b-ethylene–propylene) diblock copolymer. *RSC Advances*, **5**, 20831-20837.

12. Zebarjad, S., Bagheri, R., Reihani, S., and Lazzeri, A. (2003) Deformation, yield and fracture of elastomer-modified polypropylene. *Journal of Applied Polymer Science*, **90**, 3767-3779.

13. Wu, C. L., Zhang, M. Q., Rong, M. Z., and Friedrich, K. (2005) Silica nanoparticles filled polypropylene: effects of particle surface treatment, matrix ductility and particle species on mechanical performance of the composites. *Composites Science and Technology*, **65**, 635-645.

14. Kukaleva, N., Jollands, M., Cser, F., and Kosior, E. (2000) Influence of phase structure on impact toughening of isotactic polypropylene by metallocene-catalyzed linear low-density polyethylene. *Journal of Applied Polymer Science*, **76**, 1011-1018.

15. Xu, J., Mittal, V., and Bates, F. S. (2016) Toughened isotactic polypropylene: phase behavior and mechanical properties of blends with strategically designed random copolymer modifiers. *Macromolecules*, **49**, 6497-6506.

16. Keskin, R., and Adanur, S. (2011) Improving toughness of polypropylene with thermoplastic elastomers in injection molding. *Polymer-Plastics Technology and Engineering*, **50**, 20-28.

17. Kukackova, O., Dung, N. V., Abbrent, S., Urbanova, M., Kotek, J., and Brus, J. (2017) A novel insight into the origin of toughness in polypropylene-calcium carbonate microcomposites: Multivariate analysis of ss-NMR spectra. *Polymer*, **132**, 106-113.

18. Yang, C.-j., Huang, T., Yang, J.-h., Zhang, N., Wang, Y., and Zhou, Z.-w. (2017) Carbon nanotubes induced brittle-ductile transition behavior of the polypropylene/ethylene-propylene-diene terpolymer blends. *Composites Science and Technology*, **139**, 109-116.

19. Zhu, L., Fan, H.-N., Yang, Z.-Q., and Xu, X.-H. (2010) Evaluation of phase morphology, rheological, and mechanical properties based on polypropylene toughened with poly (ethylene-co-octene). *Polymer-Plastics Technology and Engineering*, **49**, 208-217.

20. Liang, J., and Li, R. (2000) Rubber toughening in polypropylene: A review. *Journal of Applied Polymer Science*, **77**, 409-417.

21. Zuiderduin, W., Westzaan, C., Huetink, J., and Gaymans, R. (2003) Toughening of polypropylene with calcium carbonate particles. *Polymer*, **44**, 261-275.

22. Tjong, S., Bao, S., and Liang, G. (2005) Polypropyl-ene/montmorillonite nanocomposites toughened with SEBS-g-MA: Structure - property relationship. *Journal of Polymer Science, Part B: Polymer Physics*, **43**, 3112-3126.

23. Varghese, A. M., Vengatesan, M. R., Mun, S. C., Macosko, C. W., and Mittal, V. (2018) Effect of graphene on polypropylene/maleic an-hydride grafted ethylene vinyl acetate (PP/EVA-g-MA) blend: Me-chanical, thermal, morphological and rheological properties. *Indus-trial and Engineering Chemistry Research*, **57**(23), 7834-7845.

24. Parameswaranpillai, J., Joseph, G., Shinu, K., Sreejesh, P., Jose, S., Salim, N. V., and Hameed, N. (2015) The role of SEBS in tailoring the interface between the polymer matrix and exfoliated graphene nanoplatelets in hybrid composites. *Materials Chemistry and Phys-ics*, **163**, 182-189.

25. Haghnegahdar, M., Naderi, G., and Ghoreishy, M. (2017) Fracture toughness and deformation mechanism of un-vulcanized and dy-namically vulcanized polypropylene/ethylene propylene diene monomer/graphene nanocomposites. *Composites Science and Technology*, **141**, 83-98.

26. Li, R., Zhang, X., Zhao, Y., Hu, X., Zhao, X., and Wang, D. (2009) New polypropylene blends toughened by polypropylene/poly (eth-ylene-co-propylene) in-reactor alloy: Compositional and morpho-logical influence on mechanical properties. *Polymer*, **50**, 5124-5133.

27. Gahleitner, M., Tranninger, C., and Doshev, P. (2013) Heterophasic copolymers of polypropylene: Development, design principles, and future challenges. *Journal of Applied Polymer Science*, **130**, 3028-3037.

28. Hongjun, C., Xiaolie, L., Dezhu, M., Jianmin, W., and Hongsheng, T. (1999) Structure and properties of impact copolymer polypropyl-ene. I. Chain structure. *Journal of Applied Polymer Science*, **71**, 93-101.

29. Mirabella, Jr., F. M. (1993) Impact polypropylene copolymers: Fractionation and structural characterization. *Polymer*, **34**, 1729-1735.

30. Li, Y., Xu, J.-T., Dong, Q., Fu, Z.-S., and Fan, Z.-Q. (2009) Morphology of polypropylene/poly (ethylene-co-propylene) in-reactor alloys prepared by multi-stage sequential polymerization and two-stage polymerization. *Polymer*, **50**, 5134-5141.

31. Tan, H., Li, L., Chen, Z., Song, Y., and Zheng, Q. (2005) Phase mor-phology and impact toughness of impact polypropylene copoly-mer. *Polymer*, **46**, 3522-3527.

32. Zhang, C., Shangguan, Y., Chen, R., Wu, Y., Chen, F., Zheng, Q., and Hu, G. (2010) Morphology, microstructure and compatibility of impact polypropylene copolymer. *Polymer*, **51**, 4969-4977.

33. Zhu, H., Monrabal, B., Han, C. C., and Wang, D. (2008) Phase structure and crystallization behavior of polypropylene in-reactor alloys: insights from both inter-and intramolecular compositional heterogeneity. *Macromolecules*, **41**, 826-833.

34. Doshev, P., Lach, R., Lohse, G., Heuvelsland, A., Grellmann, W., and Radusch, H.-J. (2005) Fracture characteristics and deformation behavior of heterophasic ethylene–propylene copolymers as a function of the dispersed phase composition. *Polymer*, **46**, 9411-9422.

35. Starke, J., Michler, G., Grellmann, W., Seidler, S., Gahleitner, M., Fiebig, J., and Nezbedova, E. (1998) Fracture toughness of polypropylene copolymers: Influence of interparticle distance and temperature. *Polymer*, **39**, 75-82.

36. Grein, C., Bernreitner, K., Hauer, A., Gahleitner, M., and Neißl, W. (2003) Impact modified isotatic polypropylene with controlled rubber intrinsic viscosities: some new aspects about morphology and fracture. *Journal of Applied Polymer Science*, **87**, 1702-1712.

37. Han, M.-H., and Nairn, J. A. (2003) Hygrothermal aging of polyimide matrix composite laminates. *Composites, Part A: Applied Science and Manufacturing*, **34**, 979-986.

38. Larbi, S., Bensaada, R., Bilek, A., and Djebali, S. (2015) Hygrothermal ageing effect on mechanical properties of FRP laminates. *AIP Conference Proceedings*, **1653**, 020066.

39. Hu, Y., Li, X., Lang, A. W., Zhang, Y., and Nutt, S. R. (2016) Water immersion aging of polydicyclopentadiene resin and glass fiber composites. *Polymer Degradation and Stability*, **124**, 35-42.

40. Guermazi, N., Tarjem, A. B., Ksouri, I., and Ayedi, H. F. (2016) On the durability of FRP composites for aircraft structures in hygrothermal conditioning. *Composites, Part B: Engineering*, **85**, 294-304.

41. Retegi, A., Arbelaiz, A., Alvarez, P., Llano-Ponte, R., Labidi, J., and Mondragon, I. (2006) Effects of hygrothermal ageing on mechanical properties of flax pulps and their polypropylene matrix composites. *Journal of Applied Polymer Science*, **102**, 3438-3445.

42. Mourad, A.-H., Fouad, H., and Elleithy, R. (2009) Impact of some environmental conditions on the tensile, creep-recovery, relaxation, melting and crystallinity behaviour of UHMWPE-GUR 410-medical grade. *Materials and Design*, **30**, 4112-4119.

43. He, W., Liu, N., Chen, X., Guo, J., and Wei, T. (2016) The influence of hygrothermal ageing on the mechanical properties and thermal degradation kinetics of long glass fibre reinforced polyamide 6 composites filled with sepiolite. *RSC Advances*, **6**, 36689-36697.

44. Le Gac, P.-Y., Arhant, M., Le Gall, M., and Davies, P. (2017) Yield stress changes induced by water in polyamide 6: characterization and modeling. *Polymer Degradation and Stability*, **137**, 272-280.

45. Ishak, Z. M., Ishiaku, U., and Karger-Kocsis, J. (2000) Hygrothermal

aging and fracture behavior of short-glass-fiber-reinforced rubber-toughened poly (butylene terephthalate) composites. *Composites Science and Technology*, **60**, 803-815.

46. Chen, Y., Davalos, J. F., Ray, I., and Kim, H.-Y. (2007) Accelerated aging tests for evaluations of durability performance of FRP reinforcing bars for concrete structures. *Composite Structures*, **78**, 101-111.

47. Wang, M., Xu, X., Ji, J., Yang, Y., Shen, J., and Ye, M. (2016) The hygrothermal aging process and mechanism of the novolac epoxy resin. *Composites, Part B: Engineering*, **107**, 1-8.

48. Islam, M. S., Pickering, K. L., and Foreman, N. J. (2010) Influence of hygrothermal ageing on the physico-mechanical properties of alkali treated industrial hemp fibre reinforced polylactic acid composites. *Journal of Polymers and the Environment*, **18**, 696-704.

49. Arhant, M., Le Gac, P.-Y., Le Gall, M., Burtin, C., Briançon, C., and Davies, P. (2016) Effect of sea water and humidity on the tensile and compressive properties of carbon-polyamide 6 laminates. *Composites, Part A: Applied Science and Manufacturing*, **91**, 250-261.

50. Silva, L., Tognana, S., and Salgueiro, W. (2013) Study of the water absorption and its influence on the Young's modulus in a commercial polyamide. *Polymer Testing*, **32**, 158-164.

51. Le Gac, P.-Y., Choqueuse, D., Paris, M., Recher, G., Zimmer, C., and Melot, D. (2013) Durability of polydicyclopentadiene under high temperature, high pressure and seawater (offshore oil production conditions). *Polymer Degradation and Stability*, **98**, 809-817.

52. Guadagno, L., Fontanella, C., Vittoria, V., and Longo, P. (1999) Physical aging of syndiotactic polypropylene. *Journal of Polymer Science, Part B: Polymer Physics*, **37**, 173-180.

53. Guermazi, N., Elleuch, K., Ayedi, H., and Kapsa, P. (2008) Aging effect on thermal, mechanical and tribological behaviour of polymeric coatings used for pipeline application. *Journal of Materials Processing Technology*, **203**, 404-410.

54. Gautier, L., Mortaigne, B., Bellenger, V., and Verdu, J. (2000) Osmotic cracking nucleation in hydrothermal-aged polyester matrix. *Polymer*, **41**, 2481-2490.

55. Berthé, V., Ferry, L., Bénézet, J., and Bergeret, A. (2010) Ageing of different biodegradable polyesters blends mechanical and hygrothermal behavior. *Polymer Degradation and Stability*, **95**, 262-269.

56. Boubakri, A., Haddar, N., Elleuch, K., and Bienvenu, Y. (2010) Impact of aging conditions on mechanical properties of thermoplastic polyurethane. *Materials and Design*, **31**, 4194-4201.

57. Mondal, S., Hu, J., and Yong, Z. (2006) Free volume and water vapor permeability of dense segmented polyurethane membrane. *Journal of Membrane Science*, **280**, 427-432.

58. Deshmane, C., Yuan, Q., Perkins, R., and Misra, R. (2007) On striking

variation in impact toughness of polyethylene–clay and polypro-pylene–clay nanocomposite systems: the effect of clay–polymer interaction. *Materials Science and Engineering A*, **458**, 150-157.

59. Hagemann, H., Snyder, R., Peacock, A., and Mandelkern, L. (1989) Quantitative infrared methods for the measurement of crystallinity and its temperature dependence: polyethylene. *Macromolecules*, **22**, 3600-3606.

60. Liang, C., and Krimm, S. (1959) Infrared spectra of high polymers: Part IX. Polyethylene terephthalate. *Journal of Molecular Spectroscopy*, **3**, 554-574.

61. Krimm, S., Liang, C., and Sutherland, G. (1956) Infrared spectra of high polymers. II. Polyethylene. *The Journal of Chemical Physics*, **25**, 549-562.

62. Ivanova, K. I., Pethrick, R. A., and Affrossman, S. (2001) Hygrothermal aging of rubber modified and mineral filled dicyandiamide cured digylcidyl ether of bisphenol A epoxy resin. I. Diffusion behavior. *Journal of Applied Polymer Science*, **82**, 3468-3476.

63. DeNeve, B., and Shanahan, M. E. R. (1993) Water absorption by an epoxy resin and its effect on the mechanical properties and infrared spectra. *Polymer*, **34**(24), 5099-5105.

64. Zhao, S., Chen, F., Zhao, C., Huang, Y., Dong, J.-Y., and Han, C. C. (2013) Interpenetrating network formation in isotactic polypropylene/graphene composites. *Polymer*, **54**, 3680-3690.

65. Li, C.-Q., Zha, J.-W., Long, H.-Q., Wang, S.-J., Zhang, D.-L., and Dang, Z.-M. (2017) Mechanical and dielectric properties of graphene incorporated polypropylene nanocomposites using polypropylene-graft-maleic anhydride as a compatibilizer. *Composites Science and Technology*, **153**, 111-118.

66. Chow, C., Xing, X., and Li, R. (2007) Moisture absorption studies of sisal fibre reinforced polypropylene composites. *Composites Science and Technology*, **67**, 306-313.

67. Chu, W., Wu, L., and Karbhari, V. M. (2004) Durability evaluation of moderate temperature cured E-glass/vinylester systems. *Composite Structures*, **66**, 367-376.

68. Boubakri, A., Elleuch, K., Guermazi, N., and Ayedi, H. (2009) Investigations on hygrothermal aging of thermoplastic polyurethane material. *Materials and Design*, **30**, 3958-3965.

69. Zanni-Deffarges, M., and Shanahan, M. (1995) Diffusion of water into an epoxy adhesive: comparison between bulk behaviour and adhesive joints. *International Journal of Adhesion and Adhesives*, **15**, 137-142.

70. Guermazi, N., Elleuch, K., and Ayedi, H. (2009) The effect of time and aging temperature on structural and mechanical properties of pipeline coating. *Materials and Design*, **30**, 2006-2010.

8

Effect of Accelerated UV Weathering on the Properties of Isotactic Polypropylene and Impact Polypropylene Copolymers

8.1 Introduction

Owing to tunable properties, light weight structures, ease of processing and low cost, polymer based materials have found widespread application in engineering, non-engineering and specialty fields [1-4]. In their lifespan, the polymeric materials have to sustain highly active environmental conditions, which can have detrimental effects on their performance. In brief, continuously exposing the polymeric materials to severe conditions like sunlight, UV radiation, temperature, pressure, humidity, oxygen, pollutants, chemicals, mechanical force, biological media, marine conditions, etc., results in the molecular degradation and irreversible negative effects on the physical and chemical properties [5-9]. Thus, an evaluation of the material performance in such environmental conditions is very useful to estimate the life time as well as replacement or repair needs [10,11]. Accelerated ageing studies, which suitably mimic the environmental conditions, are preferred over natural ageing because of the much longer time span needed to obtain sufficient insights about the materials in the case of natural ageing [12,13].

In laboratory, the UV radiation induced accelerated ageing of the polymeric materials can be conducted with the help of fluorescent tube, xenon long arc, mercury arc, metal halide or carbon arc lamp as effective light source [10,14], with xenon long arc representing the most preferred source. When polymer based materials are in contact with a UV source, the process of photo-initiation takes place through chromophore excitation, followed by photo-degradation by undergoing either chain scission or crosslinking [14-17]. Accordingly, color fading is the first indication of ageing, with subsequent loss in physico-chemical and mechanical properties [14,18,19]. Indeed,

Anish Varghese and Vikas Mittal*, The Petroleum Institute (part of Khalifa University of Science and Technology), Abu Dhabi, UAE
*Current address: Bletchington, Wellington County, Australia

the influence of temperature, humidity, pollutants, etc., has an accelerating effect on the degree of UV radiation induced photo-degradation [4]. One of the crucial factors deciding the degree of degradation is the nature of interaction of the UV radiations with the polymer surface, which, in turn, is related to the polymer nature [4,9]. The process of chain scission is related to the amorphous part of the polymer, which results in reduced molecular weight and increased crystallinity, whereas crosslinking relates to the imperfect crystalline parts, which results in increased molecular weight with unaltered crystallinity [15,20]. The basic reactions during photo-degradation via chain scission are classified as Norrish Type I or Norrish Type II reactions [21]. Norrish Type I reaction is the photo-chemical breakdown of aldehydes or ketones and subsequent generation of active free radical sites, while Norrish Type II reaction is the excitation of the polymer-oxygen complex through the generation of carbonyl or vinyl groups [15,21,22]. Both reaction types result in the breakdown of highly entangled macromolecular structure into smaller crystallites, which correspondingly leads to reduced molecular weight, decreased mechanical properties and increased crystallinity [15,20,21,23].

Polypropylene (PP) is a versatile polyolefin thermoplastic material available in various forms like homopolymer, copolymer, reactor-blends, etc. [24-27]. PP homopolymer, isotactic polypropylene (iPP), exhibits narrow molecular weight distribution and exceptional clarity, along with balanced physical and mechanical properties, recyclability, low production cost, corrosion resistance and thermal stability [28-30]. However, iPP suffers from poor crack propagation resistance [31,32]. The addition of elastomers, thermoplastics, thermoplastic elastomers, copolymers, etc., has been reported to be beneficial for strengthening the impact properties of iPP [30,33-44]. As an alternative, impact polypropylene copolymers (IPC), also known as heterophasic copolymers, have been developed by *in-situ* copolymerization of iPP and ethylene in the polymerization reactor using Borstar or Spheripol processes [45-48]. Generally, these processes include the bulk polymerization of iPP, followed by gas phase propylene-ethylene copolymerization [49]. The structural analysis of IPC reveals the presence of three phases such as semi-crystalline iPP phase, semi-crystalline phase of ethylene-propylene block copolymer (EPBC) and amorphous phase of ethylene-propylene random copolymer (EPRC). In other words, IPC structure can be visualized as the stress concentrator EPRC phase distributed in the iPP continuous phase under the influence of compatibilizer EPBC phase [49-51]. On impact loading, IPC leads to effective micro-voids generation via crazing or shear yielding through the absorption of impact energy by EPRC, thereby, resulting in fracture-free plastic deformation [50,52]. The factors like EPRC phase size, shape and distribution in iPP phase as well as EPRC-iPP interaction have strong influence on the overall performance [53-55].

In the current study, iPP and two IPC grades were analyzed for the effect of accelerated UV weathering conditions. For this, injection molded specimens were exposed to a fluorescent light source for two different time periods, i.e., 400 h and 700 h, together with other accelerating factors.

8.2 Experimental

8.2.1 Materials

Commercial iPP grade HD915CF and two impact polypropylene copolymer grades (termed as IPC1 and IPC2) were procured from Abu Dhabi Polymers Company Limited, UAE. The reported melt flow index (MFI) values at 230 °C and 2.16 kg were 8 and 38 g/10 min for iPP and IPC1 respectively, whereas the MFI of IPC2 was 7 g/10 min at 190 °C and 2.16 kg. The density of the PP grades was measured to be in the range 900-910 kg.m^{-3}. The polymers were used for analysis as received.

8.2.2 Sample Preparation

Dumbbell and rectangular bar shaped specimens were prepared using a plunger type lab-scale injection molding machine HAAKE MiniJet PRO (Thermo Scientific, Germany). For this purpose, PP pellets were melted by applying a temperature of 180 °C for 5 min, followed by injection into a mold at a temperature of 125 °C by applying a pressure of 430 bar for 10 s. In order to prevent the backflow of the injected sample, a holding pressure of 500 bar was applied for 6 s.

8.2.3 Photo-oxidation

Artificial photo-oxidation studies were carried out in a BGD 856 UV light accelerated weathering cabinet from Buiged Laboratory Instruments (Guangzhou), equipped with 4 fluorescent UV lamps (UVA-340) oriented horizontally on both sides in accordance with ASTM D 4329. The samples were exposed to accelerated UV conditions for 400 h and 700 h. Specifically, the accelerated weathering cycle was as follows: UV exposure with an irradiance of 0.76 W.m^{-2} at 50 °C for 8 h, combined water spray and UV exposure for 0.25 h, followed by combined condensation at 50 °C and UV exposure for 1.75 h. In this cycle, UV exposure imitates the role of sunlight, condensation imitates rain and water spray imitates dew.

8.2.4 Characterization

Fourier-transform infrared (FTIR) spectrometer (BRUKER TENSOR II Series) was used for the structural analysis of unweathered and UV weathered PP samples. FTIR spectra were generated by collecting 32 scans from the thin sections of the injection molded samples in transmission mode. A diamond attenuated total reflectance (ATR) crystal was used to obtain the spectra in the 4000-400 cm^{-1} wavenumber region with a resolution of 4 cm^{-1}.

X'Pert PRO Panalytical powder diffractometer was used to study the wide angle X-ray diffraction (WAXD) of the unweathered and UV weathered PP samples. Thin sections from the injection molded samples were scanned in the 5-60° 2-theta region using Cu-Kα irradiation with a wavelength (λ) of 1.5406 Å at 45 kV, 40 mA and room temperature. A step size of 0.017° s^{-1} and a step time of 10 s were employed.

Tensile properties of the PP samples were measured at room temperature with the aid of a 50 kN load cell universal testing machine (Instron 3345, USA). The cross-head speed used for the analysis was 10 mm.min^{-1}. The reported tensile data represents the mean of values from five dumbbell shape specimens with size of 75 mm x 5 mm x 2 mm and span length of 35 mm, in accordance with ISO 527. The effect of UV weathering on the impact strength of iPP, IPC1 and IPC2 was estimated at room temperature using Resil impactor from Ceast, USA (4 J hammer energy). For the analysis, the hammer speed of 3.64 m.s^{-1} was used. The reported impact strength represents the mean of values from five un-notched rectangular bar shaped specimens with size of 80 mm x 10 mm x 4 mm, in agreement with ISO 180.

UV weathering induced changes in the thermal properties (melting temperature and enthalpy (T_m and ΔH_f), crystallization temperature and enthalpy (T_c and ΔH_c)), percentage crystallinity (X_c), etc.) were assessed by using differential scanning calorimeter (DSC) from TA instruments, USA. 3-8 mg sample weight was examined in the temperature range of (-50, +200 °C) by applying two heating and cooling cycles using a dry nitrogen flow of 50 mL.min^{-1}. Second heating-cooling cycles were used to record the calorimetric behavior of the samples.

The morphology of the impact fractured surface of the PP samples was analyzed in a scanning electron microscope (SEM) (FEI Quanta, FEG250, USA). An accelerating voltage of 10 kV was used for the analysis. Prior to analysis, gold sputter coated samples were affixed on aluminum stubs with the help of carbon conductive adhesive tape. In addition, energy dispersive X-ray analysis (EDX) for the UV weathered PP samples was carried out by focusing on the area near the surface in order to confirm the presence of photo-oxidation products. For the examination of surface changes, micrographs were also obtained using an optical microscope Olympus BX51M, Japan, operated with an ocular magnification of 10x.

8.3 Results and Discussion

Figure 8.1 presents the FTIR spectra of unweathered and UV weathered iPP, IPC1 and IPC2 samples. For unweathered iPP, the presence of FTIR characteristic bands at 2957 cm^{-1}, 2916 cm^{-1}, 2870 cm^{-1}, 2837 cm^{-1}, 2362 cm^{-1}, 1463 cm^{-1}, 1374 cm^{-1}, 999-973-844-811 cm^{-1} and 898 cm^{-1} could be assigned to asymmetric CH_3 stretching, symmetric CH_2 stretching, symmetric CH_3 stretching, symmetric CH_2 stretching, C-H bending, asymmetric CH_3 bending, symmetric CH_3 bending, tertiary methyl skeleton deformation and stretching induced perpendicular absorption respectively [56-58]. Owing to photo-oxidation and thermal oxidation reactions on the iPP surface upon continuous UV weathering, the appearance of hydroxyl band at 3660-3200 cm^{-1}, alkyne band at 2150-2025 cm^{-1}, carbonyl band at 1850-1650 cm^{-1}, vinyl band at 1650-1560 cm^{-1} and C-O-C bonds at 1085-1055 cm^{-1} was observed for 400 h UV weathered iPP sample, as shown in Figure 8.1a [18,59-65]. Carbonyl band comprises γ-lactones, peresters, esters, conjugated ketones, aldehyde and carboxylic acids as oxidation products, whereas hydroxyl band comprises alcohols, hydroperoxides and some carboxylic acid groups [60,61]. Overall, the oxidation products evolved due to the reaction of the generated macromolecular radicals and oxygen in continuous

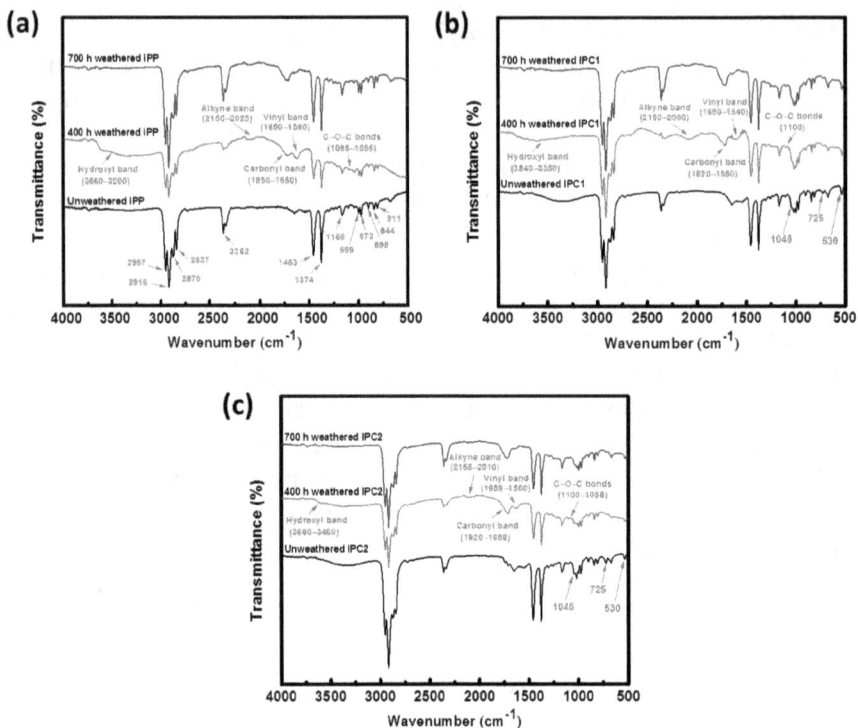

Figure 8.1 FTIR spectra of unweathered and UV weathered samples (400 h and 700 h exposure): (a) iPP, (b) IPC1 and (c) IPC2.

UV weathering conditions. For both unweathered IPC1 (Figure 8.1b) and IPC2 (Figure 8.1c), three amorphous polyethylene bands at 1045 cm-1, 725 cm-1 and 530 cm-1, assigned to CH_2 gauche configuration, CH_2 rocking and CH_2 stretching, were noticed in addition to the iPP characteristic bands [47,66-68]. Comparing the spectra of 400 h UV weathered IPC1 and IPC2 with iPP, appearance of the similar bands corresponding to the oxidation products was observed, however, the band positions differed to some extent. Accordingly, bands attributed to hydroxyl, alkyne, carbonyl and vinyl groups as well as C-O-C bonds were discerned at 3840-3350, 2150-2000, 1820-1650, 1650-1540 and 1100 cm-1 respectively for 400 h weathered IPC1 and 3660-3460, 2155-2010, 1920-1658, 1658-1560 and 1100-1058 cm-1 respectively for 400 h weathered IPC2. In addition, the intensity of the iPP, IPC1 and IPC2 bands was observed to decrease after 400 h UV weathering due to the effect of macromolecular degradation. Considering the FTIR bands of 700 h UV weathered samples, the intensity of the bands assigned to the oxidation products (except carbonyl groups) was observed to decrease probably due to the tendency to decompose [59] and/or leach under the influence of water [19]. In addition, the intensity of the iPP, IPC1 and IPC2 characteristic bands displayed an increasing trend on enhancing the exposure time from 400 to 700 h.

Figure 8.2 demonstrates the WAXD patterns of PP samples. iPP

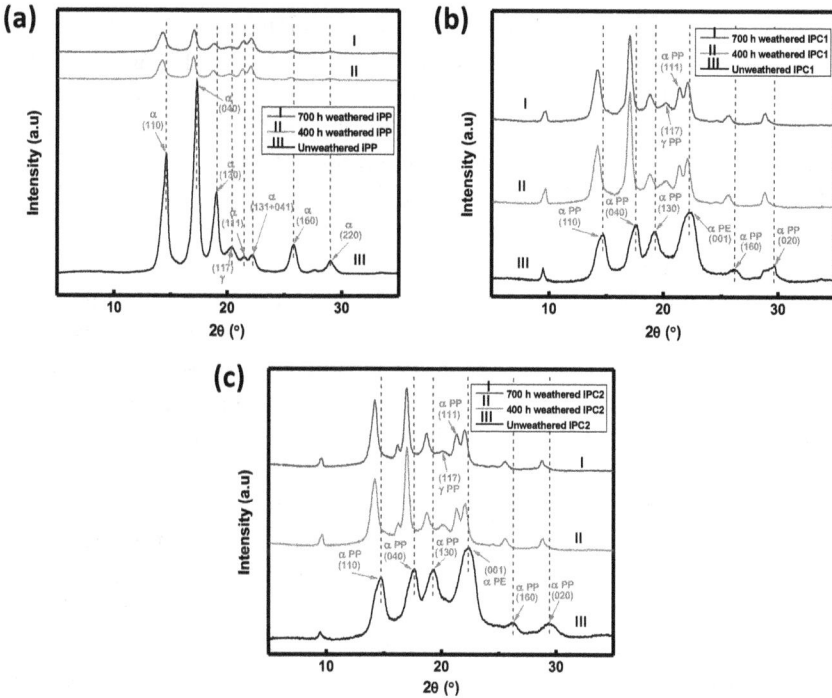

Figure 8.2 WAXD patterns of unweathered and UV weathered samples: (a) iPP, (b) IPC1 and (c) IPC2.

exhibited characteristic peaks corresponding to (110), (040), (130), (111), (131+041), (160) and (220) α-crystalline planes as well as (007) γ-crystalline plane at 14.55°, 17.26°, 19.03°, 21.47°, 22.18°, 25.76°, 29.01° and 20.41° respectively [30,69-71]. The related interplanar distance (d) was estimated by applying the Bragg's relation $d = \lambda/(2\sin\theta_{max})$, where λ (X-ray wavelength) = 1.5406 Å. As a result, the respective d values of 6.08, 5.13, 4.66, 4.14, 4.00, 3.46, 3.08 and 4.35 Å were obtained. For iPP after 400 h of UV weathering, the characteristic diffraction peaks attributed to (110), (040), (130),

(111), (131+041), (160) and (007) crystalline planes were observed to slightly shift to 14.29°, 16.97°, 18.75°, 21.39°, 22.06°, 25.62° and 20.18°, along with a reduction in the peak intensity due to significant macromolecular degradation. Accordingly, the increased d values of 6.19, 5.22, 4.73, 4.15, 4.03, 3.47 and 4.40 Å were obtained. For 700 h UV weathered iPP, the intensity of the characteristic peaks displayed an increasing trend, probably due to the chemicrystallization effect. For unweathered IPC1 and IPC2, characteristic peaks assigned to (110), (040), (130), (160) and (020) iPP α-crystalline planes as well as (001) PE α-crystalline plane were discerned at 14.67°, 17.60°, 19.19°, 26.11°, 29.69° and 22.33° respectively for IPC1 and 14.77°, 17.72°, 19.29°, 26.22°, 29.36° and 22.33° respectively for IPC2. The respective d values were 6.03, 5.04, 4.62, 3.41, 3.01 and 3.98 Å for IPC1 and 5.99, 5.00, 4.60, 3.39, 3.04 and 3.98 Å for IPC2. In the case of 400 h UV weathered IPC1, the intensity of crystalline peaks assigned to (110), (040) and (160) planes was observed to increase. It indicated the tendency of the amorphous part to undergo degradation initially, followed by the reorientation of the macromolecular structure and, thus, enhancement in crystallinity [8,17,63]. In addition, the crystalline peaks for 400 h UV weathered IPC1 shifted to 14.22°, 17.04°, 18.75°, 25.57°, 28.84° and 22.13° respectively, which further supported the degradation of the amorphous part. Interestingly, two additional peaks were also observed in the diffraction pattern at 20.19° (attributed to PP γ-crystalline plane (117)) and 21.39° (as a shoulder peak attributed to PP α-crystalline plane (111)). In the case of 700 h UV weathered IPC1, the intensity of the crystalline peaks displayed a decreasing trend indicating the macromolecular degradation. In the case of 400 and 700 h UV weathered IPC2 samples, similar behavior as IPC1 was observed. For 400 h UV weathered IPC2, three additional peaks appeared at 16.23°, 20.13° and 21.33° corresponding to PP β-crystalline plane (300), PP γ-crystalline plane (117) and PP α-crystalline plane (111) respectively. For 700 h UV weathered IPC2, the intensity of the peaks decreased because of the effect of macromolecular degradation.

Table 8.1 depicts the impact strength values of the unweathered and accelerated UV weathered samples. The impact strength values of 7, 53, and 103 kJ.m⁻² were recorded for unweathered iPP, IPC1 and IPC2 respectively (Figure 8.3a). The significantly higher impact strength for IPC1 and IPC2 resulted due to the influence of multiphase structural composition, i.e., iPP, EPRC and EPBC [50,51]. Here,

Table 8.1 Mechanical properties of the PP samples

Sample	Impact strength (kJ.m^{-2})	Tensile strength (MPa)	Tensile modulus (MPa)	Extension at break (%)
Unweathered iPP	7	39	1011	24
400 h UV weathered iPP	3	15	675	8
700 h UV weathered iPP	2.5	8	315	7
Unweathered IPC1	53	22	1018	35
400 h UV weathered IPC1	3.7	10	569	7
700 h UV weathered IPC1	3	9	495	7
Unweathered IPC2	103	21	864	299
400 h UV weathered IPC2	6.1	9	345	8
700 h UV weathered IPC2	5.9	7	219	7

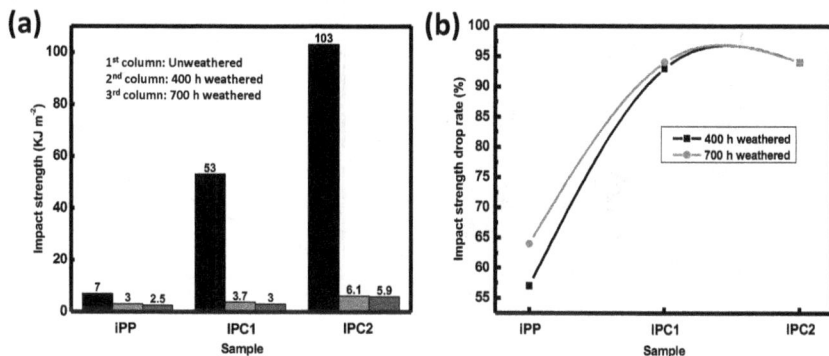

Figure 8.3 (a) Impact strength of unweathered and UV weathered samples; (b) impact strength drop rate of 400 and 700 h UV weathered samples.

EPRC can act as robust stress absorber, whereas EPBC can act as a compatibilizer to strengthen the interfacial performance between iPP and EPRC. Under impact load, EPRC can, thus, absorb a majority of the load and generate micro-voids through crazing and/or shear yielding, which provide improved stress propagation resistance [50,54]. In addition, the difference in the impact strength of IPC1 and IPC2 could have resulted due to their multiphase structural behavior, i.e., interaction between iPP and EPRC, along with distribution, size and shape of EPRC [52,53,72]. The impact strength of the samples was observed to decrease significantly after 400 h of UV weathering. For instance, ca. 57, 93 and 94% loss in the impact strength were recorded for iPP, IPC1 and IPC2 respectively (Figure 8.3b). The observed loss in the impact strength values for 700 h UV weathered samples was ca. 64, 94 and 94% respectively. Thus, a slowdown in the impact strength loss was observed on increasing the weathering time to 700 h from 400 h. The decrease in the impact strength can be attributed to the structural embrittlement resulting from the macromolecular degradation. Therefore, the weathered materials exhibited diminished capability to dissipate impact energy [17,73,74]. Also, the greater extent of impact strength loss for IPC1 and IPC2 as compared to iPP indicated the influence of amorphous part and its severe degradation in IPC1 and IPC2.

The tensile modulus of unweathered iPP, IPC1 and IPC2 was measured to be 1011, 1018 and 864 MPa respectively (Figure 8.4a).

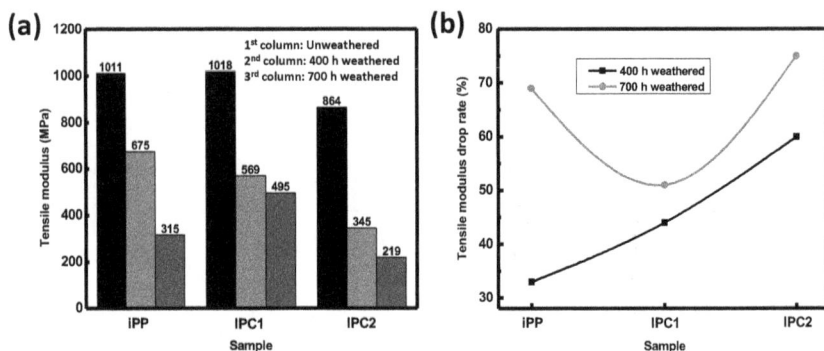

Figure 8.4 (a) Tensile modulus of unweathered and UV weathered samples; (b) tensile modulus drop rate of 400 and 700 h UV weathered samples.

The tensile modulus of iPP, IPC1 and IPC2 dropped by ca. 33, 44 and 60% respectively after 400 h UV weathering (Figure 8.4b). The observed trend of tensile modulus drop was also visible in the case of 700 h UV weathered samples, where the loss was ca. 69, 51 and 75% for iPP, IPC1 and IPC2 respectively. The significant drop in tensile modulus is probably caused by the combined effect of surface cracks and macromolecular degradation. Here, the macromolecular degradation is related to the molecular weight drop resulting due to the generation of oxidation products upon continuous UV weathering [8,75-55]. Furthermore, on enhancing the UV weathering time, the degree of surface crack formation would also increase [62,78]. Regardless of the polymer type, the PP grades also suffered a drop in the tensile strength after continuous UV weathering. For instance, ca. 62, 55 and 57% drop in the tensile strength were recorded for iPP, IPC1 and IPC2 respectively after 400 h UV weathering, whereas these values were ca. 79, 59 and 67% for 700 h UV weathered samples. Among the PP samples, iPP suffered a large tensile strength drop after 400 h and 700 h UV weathering, whereas IPC1 displayed relatively higher resistance against the loss of tensile strength. The observed significant drop in the tensile strength of the UV weathered PP samples can be correlated with the macromolecular degradation induced structural embrittlement upon continuous UV weathering [3,17,55]. Furthermore, the surface cracks are expected to have an additional effect on the tensile strength of the samples [16]. As observed for other tensile properties, the extension at break also suffered a reduction after UV weathering. For instance, after 400 h of UV weathering, the extension at break of iPP, IPC1 and IPC2 dropped by ca. 67, 80 and 97% respectively, whereas the drop was 71, 80 and 98% for 700 h UV weathered samples. Macromolecular degradation resulting from the significant drop in molecular weight can be attributed to the observed loss in elongation [78-80]. The appearance of nearly stabilized elongation in the 400-700 h UV weathering range may result due to the termination of the surface crack generation process once the photo-oxidation process has completed [8].

The calorimetric properties of the samples are presented in Table 8.2, and the second heating and cooling traces of the samples are plotted in Figures 8.5 and 8.6. The T_m values of 168.5, 166.5 and 167.7 °C were observed for unweathered iPP, IPC1 and IPC2. Due to the detrimental effect of UV weathering leading to macromolecular degradation and molecular weight reduction, the T_m values of the samples were reduced, regardless of their type [3,16,17,55,73,81]. Specifically, the T_m drop of 12.3, 6.4 and 12.6 °C was observed for iPP, IPC1 and IPC2 after 400 h UV weathering. The observed drop in T_m slowed down in the UV weathering period of 400-700 h. The T_c values of 127.9, 130.4 and 128.9 °C were observed for unweathered iPP, IPC1 and IPC2 respectively. T_c for iPP, IPC1 and IPC2 after 400 h of UV weathering was reduced by 4, 2.6 and 2.5 °C respectively. For 700 h of UV weathering, the recorded reduction in T_c was 4.5, 3 and 3.9 °C respectively. The observed drop in T_c of the PP grades indicated reduced molecular weight and an enhanced degree of chemical deformities [19,82]. Regardless of the type, ΔH_f and ΔH_c of the PP grades were observed to increase with UV weathering time, thus,

Table 8.2 Calorimetric properties of the samples

Sample	T_c (°C)	ΔT_c (°C)	ΔH_c (J/g)	T_m (°C)	ΔT_m (°C)	ΔH_f (J/g)	X_c (%)
Unweathered iPP	127.9	-	98.7	168.5	-	90.8	43.8
400 h UV weathered iPP	123.9	-4	102.1	156.2	-12.3	105.3	50.8
700 h UV weathered iPP	123.4	-4.5	107.8	154.7	-13.8	110.9	53.5
Unweathered IPC1	130.4	-	84.3	166.5	-	82.8	40.0
400 h UV weathered IPC1	127.8	-2.6	87.3	160.1	-6.4	89.9	43.3
700 h UV weathered IPC1	127.4	-3	88.2	157.8	-8.7	91.8	44.3
Unweathered IPC2	128.9	-	74.5	167.7	-	72.9	35.2
400 h UV weathered IPC2	126.4	-2.5	77.1	155.1	-12.6	79.4	38.3
700 h UV weathered IPC2	125.0	-3.9	77.9	153.2	-14.5	80.8	39.0

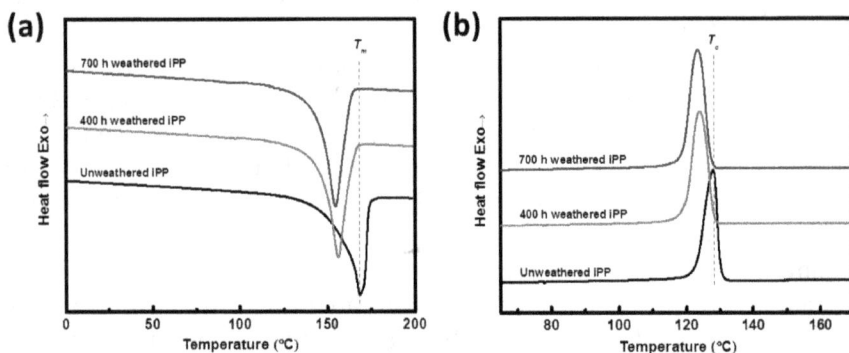

Figure 8.5 DSC (a) heating and (b) cooling thermograms of unweathered and UV weathered iPP.

indicating increased X_c. The following equation [30,43] was used for the estimation of X_c for the samples:

$$X_c = \frac{\Delta H}{\Delta H_0} \times 100$$

Here, ΔH_0 and ΔH correspond to theoretical ΔH_f of 100% crystalline PP (207.1 J/g) [30] and DSC measured ΔH_f of the samples. The

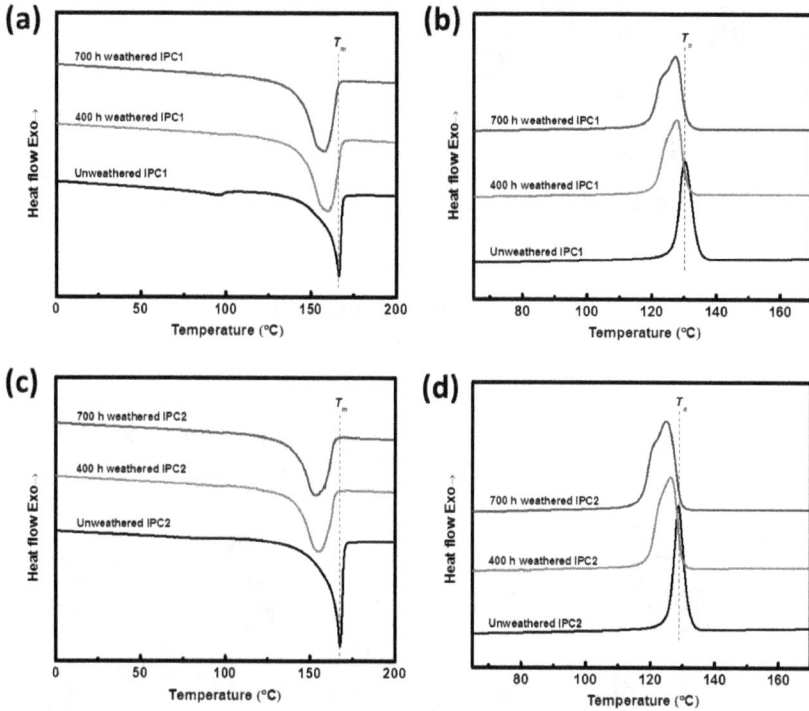

Figure 8.6 DSC heating thermograms of unweathered and UV weathered samples: (a) IPC1 and (c) IPC2; and cooling thermograms of unweathered and UV weathered samples: (b) IPC1 and (d) IPC2.

X_c values of 43.8, 40.0 and 35.2% were calculated for unweathered iPP, IPC1 and IPC2 respectively (Figure 8.7a). After continuous UV weathering, the X_c values of the PP grades were observed to increase, regardless of the type. Specifically, X_c increase of 9.7, 4.3 and 3.8% was recorded for 700 h UV weathered samples (Figure 8.7b). Such increase in X_c is opined to be the effect of chemi-crystallization which leads to the reorganization of the degraded non-crystalline sections into crystallites [8,63,73,83]. It is considered as an indirect effect of UV weathering. In accordance with such changes, the cracks would appear on the sample surface due to shrinkage and associated dimensional changes [17,84].

The analysis of the UV weathering induced changes in the micromorphology of the impact fractured surface of iPP, IPC1 and IPC2

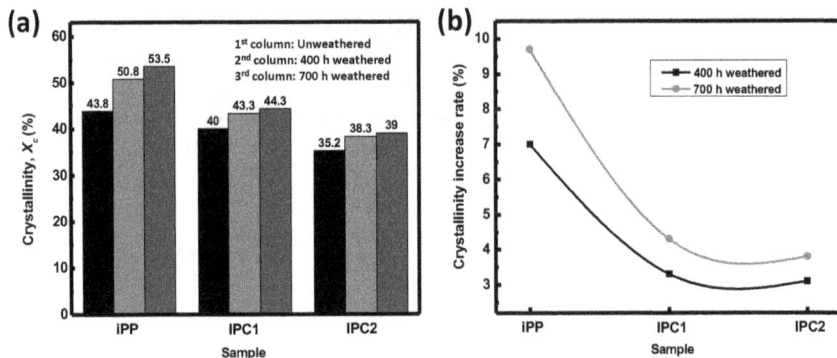

Figure 8.7 (a) Percentage crystallinity of unweathered and UV weathered samples; (b) percentage crystallinity increase rate of 400 and 700 h UV weathered samples.

was performed using SEM, as demonstrated in Figure 8.8. The fracture morphology of iPP indicated an enhancement in the brittle

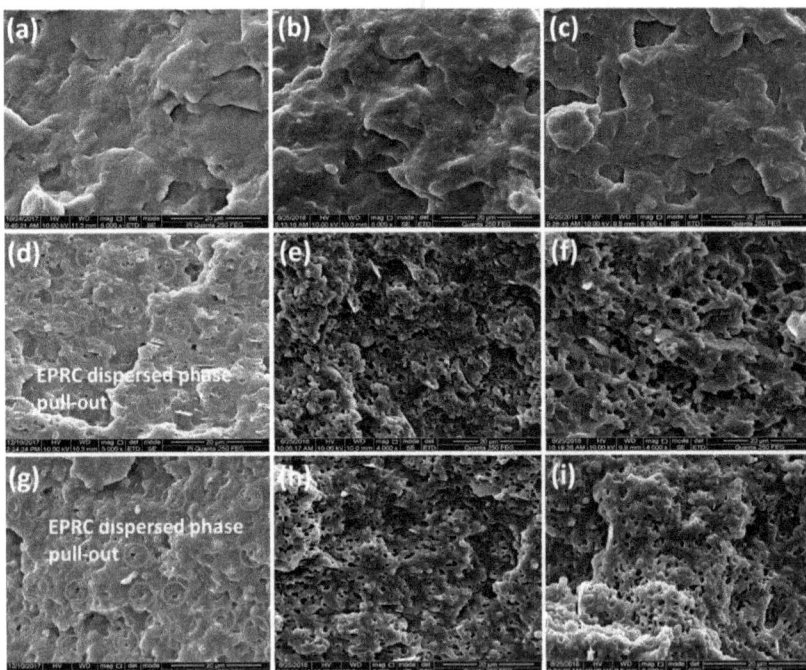

Figure 8.8 Micrographs of the impact fractured surface of unweathered (a) iPP, (d) IPC1 and (g) IPC2; 400 h UV weathered (b) iPP, (e) IPC1 and (h) IPC2; and 700 h UV weathered (c) iPP, (f) IPC1 and (i) IPC2.

nature with UV weathering time. The observed structural embrit-tlement can be expected to result from the photo-oxidation induced macromolecular degradation during UV weathering. These results coincided well with the impact properties of the unweathered and UV weathered iPP samples. Due to increased structural embrittle-ment and corresponding impact load dissipation inability, the im-pact strength values dropped for the UV weathered iPP samples. In the micrographs for IPC1 and IPC2, the observed micro-voids or dispersed EPRC phase pull-outs in the continuous phase evidenced the successful iPP-EPRC interface debonding or cavitation under impact loading, by means of either shear yielding or crazing. Also, the observed higher impact strength of IPC2 can be attributed to the influence of more desirable interfacial features between iPP and EPRC in the presence of EPBC. UV weathered IPC1 and IPC2 exhibit-ed severe damage to the EPRC dispersed phase and iPP-EPRC inter-face. These observations underline the photo-degradation of the amorphous regions present in the IPC macromolecular structure and the corresponding increase in embrittlement when exposed to continuous UV weathering.

The EDX spectra in Figure 8.9 represent the elemental composi-tion with respect to carbon (C) and oxygen (O) content. Table 8.3 also lists the recorded C and O content as well as the C/O ratio for 400 h and 700 h UV weathered iPP, IPC1 and IPC2. An increasing O content with weathering time was observed, which supported the evolution of photo-oxidation products. Specifically, enhanced O con-tent of 18.86, 23.21 and 24.44 wt% was recorded for 700 h UV weathered iPP, IPC1 and IPC2 respectively. Correspondingly, the C/O ratio exhibited a decreasing trend with UV weathering time. After 700 h of UV weathering, the observed C/O ratio for iPP, IPC1 and IPC2 samples was 4.30, 3.31 and 3.09 respectively. Overall, the results indicated the presence of more oxidation products in the UV weathered IPC2 as compared to iPP.

The surface morphology of the UV weathered samples was also analyzed through optical analysis. The optical micrographs of 400 h and 700 h UV weathered iPP, IPC1 and IPC2 samples are depicted in Figure 8.10. UV weathering induced surface degradation was ob-served as surface cracks in the PP samples, regardless of their type. Specifically, the appearance of a transverse crack network was no-ticed on the UV weathered surface of the samples, which became more intense on increasing the weathering time. Careful analysis also revealed that the nature of the surface cracks was different for

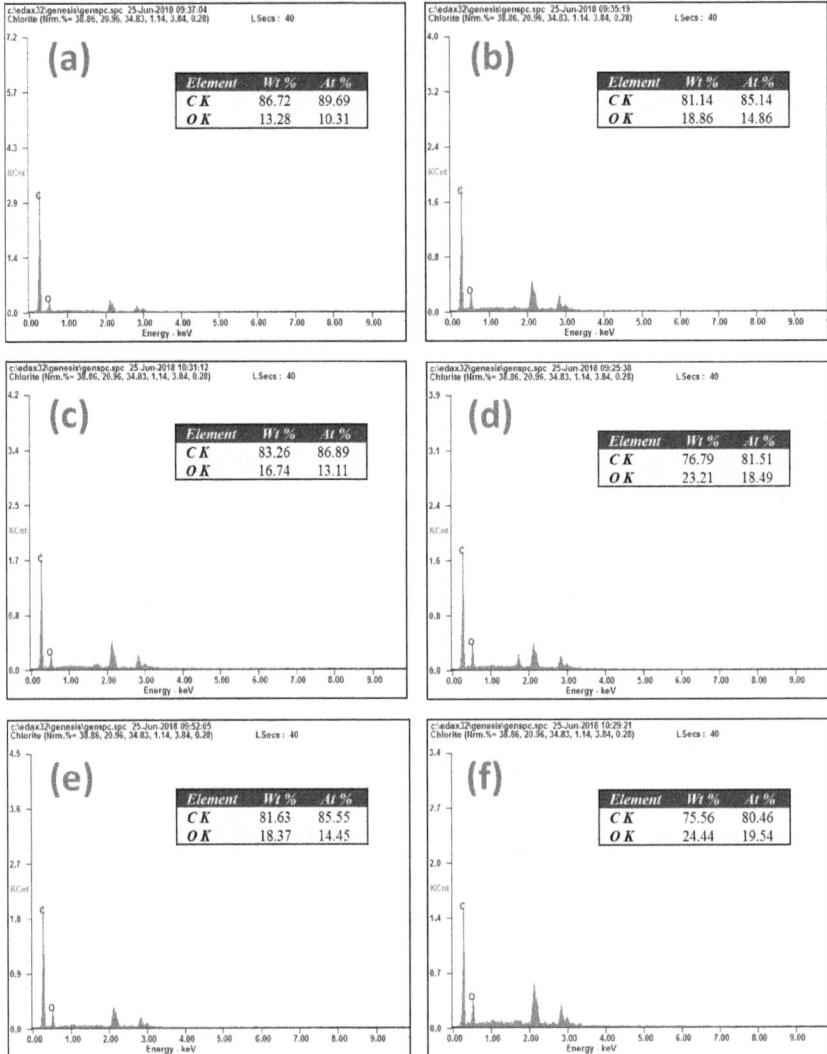

Figure 8.9 EDX spectra of 400 h UV weathered samples: (a) iPP, (c) IPC1 and (e) IPC2; and 700 h UV weathered samples: (b) iPP, (d) IPC1 and (f) IPC2.

each PP grade. For iPP, transverse cracks with smaller width were observed in large numbers after 700 h of UV weathering, whereas no surface crack was visible after 400 h. Large width transverse cracks in small numbers were observed for IPC1 after 400 h and 700 h of weathering in UV conditions. In the case of IPC2, medium width slightly tapered transverse cracks in large numbers were observed.

Table 8.3 EDX data for unweathered and UV weathered iPP, IPC1 and IPC2

Sample	Carbon (C)		Oxygen (O)		C/O ratio	
	Wt%	At%	Wt%	At%	Wt%	At%
400 h weathered iPP	86.72	89.69	13.28	10.31	6.53	8.70
700 h weathered iPP	81.14	85.14	18.86	14.86	4.30	5.73
400 h weathered IPC1	83.26	86.89	16.74	13.11	4.97	6.63
700 h weathered IPC1	76.79	81.51	23.21	18.49	3.31	4.41
400 h weathered IPC2	81.63	85.55	18.37	14.45	4.44	5.92
700 h weathered IPC2	75.56	80.46	24.44	19.54	3.09	4.12

The appearance of such surface cracks during continuous UV weathering is related to shrinkage and associated dimensional change resulting from chemi-crystallization of the photo-degraded non-crystalline sections [17,84].

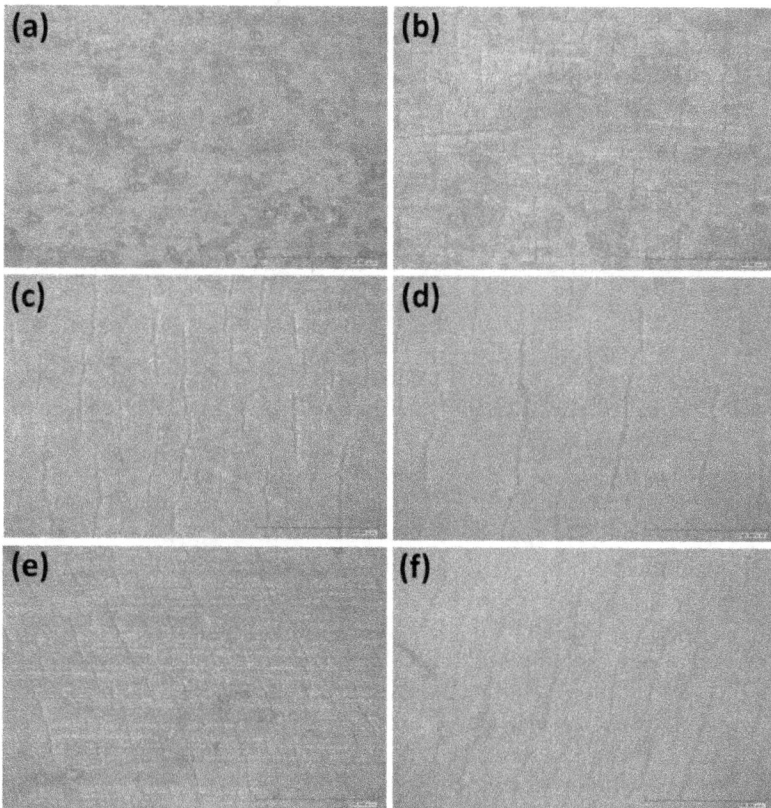

Figure 8.10 Surface micrographs of 400 h UV weathered (a) iPP, (c) IPC1 and (e) IPC2; and 700 h UV weathered (b) iPP, (d) IPC1 and (f) IPC2.

8.4 Conclusions

The objective of the study was to compare the photo-degradation of isotactic polypropylene (iPP) and impact polypropylene copolymers (IPC1 and IPC2) in continuous UV weathering conditions. The FTIR spectra of the unweathered and UV weathered samples indicated the formation of various oxidation products upon UV weathering. These observations were further supported by the EDX results associated with the elemental composition. Due to the combined effect of macromolecular degradation, structural embrittlement and surface cracks, the mechanical properties such as impact strength, tensile modulus, tensile strength and extension at break were significantly reduced. Similar to the mechanical properties, melting and crystallization temperatures, along with melting and crystallization enthalpy, were also decreased after UV weathering due to the aforementioned weathering induced changes. Regardless of the type, the crystallinity of the PP samples increased with UV weathering due to the chemi-crystallization phenomenon.

References

1. Brinson, H. F., and Brinson, L. C. (2015). Characteristics, applications and properties of polymers. In: *Polymer Engineering Science and Viscoelasticity*, Springer, USA, pp. 57-100.
2. Paoletti, A., Lambiase, F., and Di Ilio, A. (2016) Analysis of forces and temperatures in friction spot stir welding of thermoplastic polymers. *The International Journal of Advanced Manufacturing Technology*, **83**, 1395-1407.
3. Darie, R. N., Bodirlau, R., Teaca, C. A., Macyszyn, J., Kozlowski, M., and Spiridon, I. (2013) Influence of accelerated weathering on the properties of polypropylene/polylactic acid/eucalyptus wood composites. *International Journal of Polymer Analysis and Characterization*, **18**, 315-327.
4. Pandey, J. K., Reddy, K. R., Kumar, A. P., and Singh, R. (2005) An overview on the degradability of polymer nanocomposites. *Polymer Degradation and Stability*, **88**, 234-250.
5. Pickett, J., Gibson, D., and Gardner, M. (2008) Effects of irradiation conditions on the weathering of engineering thermoplastics. *Polymer Degradation and Stability*, **93**, 1597-1606.
6. Tan, K., White, C., Benatti, D., and Hunston, D. (2010) Effects of ultraviolet radiation, temperature and moisture on aging of coatings and sealants–A chemical and rheological study. *Polymer Degrada-*

tion and Stability*, **95**, 1551-1556.

7. Tan, K., White, C., Benatti, D., and Hunston, D. (2008) Evaluating aging of coatings and sealants: Mechanisms. *Polymer Degradation and Stability*, **93**, 648-656.

8. Yakimets, I., Lai, D., and Guigon, M. (2004) Effect of photo-oxidation cracks on behaviour of thick polypropylene samples. *Polymer Degradation and Stability*, **86**, 59-67.

9. Kumar, A. P., Depan, D., Tomer, N. S., and Singh, R. P. (2009) Nanoscale particles for polymer degradation and stabilization - trends and future perspectives. *Progress in Polymer Science*, **34**, 479-515.

10. Heikkilä, A., Kärhä, P., Tanskanen, A., Kaunismaa, M., Koskela, T., Kaurola, J., Ture, T., and Syrjälä, S. (2009) Characterizing a UV chamber with mercury lamps for assessment of comparability to natural UV conditions. *Polymer Testing*, **28**, 57-65.

11. Larché, J.-F., Bussière, P.-O., Therias, S., and Gardette, J.-L. (2012) Photooxidation of polymers: Relating material properties to chemical changes. *Polymer Degradation and Stability*, **97**, 25-34.

12. Azwa, Z., Yousif, B., Manalo, A., and Karunasena, W. (2013) A review on the degradability of polymeric composites based on natural fibres. *Materials & Design*, **47**, 424-442.

13. Chen, Y., Davalos, J. F., Ray, I., and Kim, H.-Y. (2007) Accelerated aging tests for evaluations of durability performance of FRP reinforcing bars for concrete structures. *Composite Structures*, **78**, 101-111.

14. Tocháček, J., and Vrátníčková, Z. (2014) Polymer life-time prediction: The role of temperature in UV accelerated ageing of polypropylene and its copolymers. *Polymer Testing*, **36**, 82-87.

15. Stark, N. M., and Matuana, L. M. (2004) Surface chemistry changes of weathered HDPE/wood-flour composites studied by XPS and FTIR spectroscopy. *Polymer Degradation and Stability*, **86**, 1-9.

16. Rabello, M., and White, J. (1997) The role of physical structure and morphology in the photodegradation behaviour of polypropylene. *Polymer Degradation and Stability*, **56**, 55-73.

17. Beg, M. D. H., and Pickering, K. L. (2008) Accelerated weathering of unbleached and bleached Kraft wood fibre reinforced polypropylene composites. *Polymer Degradation and Stability*, **93**, 1939-1946.

18. La Mantia, F., and Morreale, M. (2008) Accelerated weathering of polypropylene/wood flour composites. *Polymer Degradation and Stability*, **93**, 1252-1258.

19. Butylina, S., Hyvärinen, M., and Kärki, T. (2012) A study of surface changes of wood-polypropylene composites as the result of exterior weathering. *Polymer Degradation and Stability*, **97**, 337-345.

20. Jabarin, S. A., and Lofgren, E. A. (1994) Photooxidative effects on properties and structure of high-density polyethylene. *Journal of*

Applied Polymer Science, **53**, 411-423.

21. Fabiyi, J. S., McDonald, A. G., Wolcott, M. P., and Griffiths, P. R. (2008) Wood plastic composites weathering: Visual appearance and chemical changes. *Polymer Degradation and Stability*, **93**, 1405-1414.

22. *Handbook of Polymer Degradation*, Hamid, S. H. (ed.), CRC Press, USA (2000).

23. Philip, M., Attwood, J., Hulme, A., Williams, G., and Shipton, P. (2004) Evaluation of Weathering in Mixed Polyethylene and Polypropylene Products, The Waste and Resources Action Program, UK. Online: https://cms.esi.info/Media/documents/11653_13615 34798115.pdf [accessed 19th May 2019].

24. Tripathi, D. (2002) *Practical Guide to Polypropylene*, iSmithers Rapra Publishing, USA.

25. *Handbook of Polypropylene and Polypropylene Composites*, Karian, H. G. (ed.), CRC Press, USA (2003).

26. Maddah, H. A. (2016) Polypropylene as a promising plastic: A review. *American Journal of Polymer Science*, **6**, 1-11.

27. Gahleitner, M., and Paulik, C. (2017) Polypropylene and other polyolefins. In: *Brydson's Plastics Materials*, 8th edition, Elsevier, USA, pp. 279-309.

28. Bikiaris, D., Vassiliou, A., Chrissafis, K., Paraskevopoulos, K., Jannakoudakis, A., and Docoslis, A. (2008) Effect of acid treated multiwalled carbon nanotubes on the mechanical, permeability, thermal properties and thermo-oxidative stability of isotactic polypropylene. *Polymer Degradation and Stability*, **93**, 952-967.

29. Wang, Y., and Tsai, H.B. (2012) Thermal, dynamic-mechanical, and dielectric properties of surfactant intercalated graphite oxide filled maleated polypropylene nanocomposites. *Journal of Applied Polymer Science*, **123**, 3154-3163.

30. Varghese, A. M., Vengatesan, M. R., Mun, S. C., Macosko, C. W., and Mittal, V. (2018) Effect of graphene on polypropylene/maleic anhydride-graft-ethylene-vinyl acetate (PP/EVA-g-MA) blend: Mechanical, thermal, morphological, and rheological properties. *Industrial & Engineering Chemistry Research*, **57**, 7834-7845.

31. Chen, F., Qiu, B., Wang, B., Shangguan, Y., and Zheng, Q. (2015) Balanced toughening and strengthening of ethylene–propylene rubber toughened isotactic polypropylene using a poly (styrene-b-ethylene–propylene) diblock copolymer. *RSC Advances*, **5**, 20831-20837.

32. Zebarjad, S., Bagheri, R., Reihani, S.S., and Lazzeri, A. (2003) Deformation, yield and fracture of elastomer-modified polypropylene. *Journal of Applied Polymer Science*, **90**, 3767-3779.

33. Liang, J., and Li, R. (2000) Rubber toughening in polypropylene: A review. *Journal of Applied Polymer Science*, **77**, 409-417.

34. Kukaleva, N., Jollands, M., Cser, F., and Kosior, E. (2000) Influence of phase structure on impact toughening of isotactic polypropylene by metallocene-catalyzed linear low-density polyethylene. *Journal of Applied Polymer Science*, **76**, 1011-1018.
35. Keskin, R., and Adanur, S. (2011) Improving toughness of polypropylene with thermoplastic elastomers in injection molding. *Polymer-Plastics Technology and Engineering*, **50**, 20-28.
36. Zhu, L., Fan, H.-N., Yang, Z.-Q., and Xu, X.-H. (2010) Evaluation of phase morphology, rheological, and mechanical properties based on polypropylene toughened with poly (ethylene-co-octene). *Polymer-Plastics Technology and Engineering*, **49**, 208-217.
37. Xu, J., Mittal, V., and Bates, F. S. (2016) Toughened isotactic polypropylene: phase behavior and mechanical properties of blends with strategically designed random copolymer modifiers. *Macromolecules*, **49**, 6497-6506.
38. Wu, C. L., Zhang, M. Q., Rong, M. Z., and Friedrich, K. (2005) Silica nanoparticles filled polypropylene: effects of particle surface treatment, matrix ductility and particle species on mechanical performance of the composites. *Composites Science And Technology*, **65**, 635-645.
39. Zuiderduin, W., Westzaan, C., Huetink, J., and Gaymans, R. (2003) Toughening of polypropylene with calcium carbonate particles. *Polymer*, **44**, 261-275.
40. Kukackova, O., Dung, N. V., Abbrent, S., Urbanova, M., Kotek, J., and Brus, J. (2017) A novel insight into the origin of toughness in polypropylene–calcium carbonate microcomposites: Multivariate analysis of ss-NMR spectra. *Polymer*, **132**, 106-113.
41. Yang, C.-j., Huang, T., Yang, J.-h., Zhang, N., Wang, Y., and Zhou, Z.-w. (2017) Carbon nanotubes induced brittle-ductile transition behavior of the polypropylene/ethylene-propylene-diene terpolymer blends. *Composites Science and Technology*, **139**, 109-116.
42. Tjong, S., Bao, S., and Liang, G. (2005) Polypropylene/montmorillonite nanocomposites toughened with SEBS-g-MA: Structure–property relationship. *Journal of Polymer Science, Part B: Polymer Physics*, **43**, 3112-3126.
43. Parameswaranpillai, J., Joseph, G., Shinu, K., Sreejesh, P., Jose, S., Salim, N. V., and Hameed, N. (2015) The role of SEBS in tailoring the interface between the polymer matrix and exfoliated graphene nanoplatelets in hybrid composites. *Materials Chemistry and Physics*, **163**, 182-189.
44. Haghnegahdar, M., Naderi, G., and Ghoreishy, M. (2017) Fracture toughness and deformation mechanism of un-vulcanized and dynamically vulcanized polypropylene/ethylene propylene diene monomer/graphene nanocomposites. *Composites Science and Technology*, **141**, 83-98.

45. Li, R., Zhang, X., Zhao, Y., Hu, X., Zhao, X., and Wang, D. (2009) New polypropylene blends toughened by polypropylene/poly (ethylene-co-propylene) in-reactor alloy: Compositional and morphological influence on mechanical properties. *Polymer*, **50**, 5124-5133.

46. Gahleitner, M., Tranninger, C., and Doshev, P. (2013) Heterophasic copolymers of polypropylene: Development, design principles, and future challenges. *Journal of Applied Polymer Science*, **130**, 3028-3037.

47. Hongjun, C., Xiaolie, L., Dezhu, M., Jianmin, W., and Hongsheng, T. (1999) Structure and properties of impact copolymer polypropylene. I. Chain structure. *Journal of Applied Polymer Science*, **71**, 93-101.

48. Mirabella, Jr., F. M. (1993) Impact polypropylene copolymers: fractionation and structural characterization. *Polymer*, **34**, 1729-1735.

49. Li, Y., Xu, J.-T., Dong, Q., Fu, Z.-S., and Fan, Z.-Q. (2009) Morphology of polypropylene/poly (ethylene-co-propylene) in-reactor alloys prepared by multi-stage sequential polymerization and two-stage polymerization. *Polymer*, **50**, 5134-5141.

50. Zhang, C., Shangguan, Y., Chen, R., Wu, Y., Chen, F., Zheng, Q., and Hu, G. (2010) Morphology, microstructure and compatibility of impact polypropylene copolymer. *Polymer*, **51**, 4969-4977.

51. Tan, H., Li, L., Chen, Z., Song, Y., and Zheng, Q. (2005) Phase morphology and impact toughness of impact polypropylene copolymer. *Polymer*, **46**, 3522-3527.

52. Grein, C., Bernreitner, K., Hauer, A., Gahleitner, M., and Neißl, W. (2003) Impact modified isotatic polypropylene with controlled rubber intrinsic viscosities: Some new aspects about morphology and fracture. *Journal of Applied Polymer Science*, **87**, 1702-1712.

53. Doshev, P., Lach, R., Lohse, G., Heuvelsland, A., Grellmann, W., and Radusch, H.-J. (2005) Fracture characteristics and deformation behavior of heterophasic ethylene–propylene copolymers as a function of the dispersed phase composition. *Polymer*, **46**, 9411-9422.

54. Grellmann, W., Seidler, S., Gahleitner, M., and Nezbedova, E. (1998) Fracture toughness of polypropylene copolymers: influence of interparticle distance and temperature. *Polymer*, **39**, 75-82.

55. Joseph, P., Rabello, M. S., Mattoso, L., Joseph, K., and Thomas, S. (2002) Environmental effects on the degradation behaviour of sisal fibre reinforced polypropylene composites. *Composites Science and Technology*, **62**, 1357-1372.

56. Deshmane, C., Yuan, Q., Perkins, R., and Misra, R. (2007) On striking variation in impact toughness of polyethylene–clay and polypropylene–clay nanocomposite systems: the effect of clay–polymer interaction. *Materials Science and Engineering A*, **458**, 150-157.

57. Wu, Q., and Qu, B. (2001) Synergistic effects of silicotungistic acid

on intumescent flame-retardant polypropylene. *Polymer Degradation and Stability*, **74**, 255-261.

58. Butylina, S., Hyvärinen, M., and Kärki, T. (2012) Accelerated weathering of wood–polypropylene composites containing minerals. *Composites, Part A: Applied Science and Manufacturing*, **43**, 2087-2094.

59. Lv, Y., Huang, Y., Yang, J., Kong, M., Yang, H., Zhao, J., and Li, G. (2015) Outdoor and accelerated laboratory weathering of polypropylene: A comparison and correlation study. *Polymer Degradation and Stability*, **112**, 145-159.

60. Bocchini, S., Morlat-Therias, S., Gardette, J. L., and Camino, G. (2008) Influence of nanodispersed hydrotalcite on polypropylene photooxidation. *European Polymer Journal*, **44**, 3473-3481.

61. Gadioli, R., Morais, J. A., Waldman, W. R., and De Paoli, M. -A. (2014) The role of lignin in polypropylene composites with semi-bleached cellulose fibers: mechanical properties and its activity as antioxidant. *Polymer Degradation and Stability*, **108**, 23-34.

62. Goel, A., Chawla, K., Vaidya, U., Koopman, M., and Dean, D. (2008) Effect of UV exposure on the microstructure and mechanical properties of long fiber thermoplastic (LFT) composites. *Journal of Materials Science*, **43**, 4423-4432.

63. Grigoriadou, I., Paraskevopoulos, K., Karavasili, M., Karagiannis, G., Vasileiou, A., and Bikiaris, D. (2013) HDPE/Cu-nanofiber nanocomposites with enhanced mechanical and UV stability properties. *Composites, Part B: Engineering*, **55**, 407-420.

64. Guadagno, L., Naddeo, C., and Vittoria, V. (2004) Structural and morphological changes during UV irradiation of the crystalline helical form of syndiotactic polypropylene. *Macromolecules*, **37**, 9826-9834.

65. Vallejo-Montesinos, J., Martínez, J. L., Montejano-Carrizales, J., Perez, E., Pérez, J. B., Almendárez-Camarillo, A., and Gonzalez-Calderon, J. (2017) Passivation of titanium oxide in polyethylene matrices using polyelectrolytes as titanium dioxide surface coating. *Mechanics, Materials Science & Engineering Journal*, **8**, doi: 10.2412/mmse.96.48.950.

66. Krimm, S., Liang, C., and Sutherland, G. (1956) Infrared spectra of high polymers. II. Polyethylene. *The Journal of Chemical Physics*, **25**, 549-562.

67. Hagemann, H., Snyder, R., Peacock, A., and Mandelkern, L. (1989) Quantitative infrared methods for the measurement of crystallinity and its temperature dependence: polyethylene. *Macromolecules*, **22**, 3600-3606.

68. Liang, C., and Krimm, S. (1959) Infrared spectra of high polymers: Part IX. Polyethylene terephthalate. *Journal of Molecular Spectroscopy*, **3**, 554-574.

69. Li, C.-Q., Zha, J.-W., Long, H.-Q., Wang, S.-J., Zhang, D.-L., and Dang, Z.-M. (2017) Mechanical and dielectric properties of graphene incorporated polypropylene nanocomposites using polypropylene-graft-maleic anhydride as a compatibilizer. *Composites Science and Technology*, **153**, 111-118.

70. Zhao, S., Chen, F., Zhao, C., Huang, Y., Dong, J.-Y., and Han, C. C. (2013) Interpenetrating network formation in isotactic polypropylene/graphene composites. *Polymer*, **54**, 3680-3690.

71. Yang, C., Wang, M., Xing, Z., Zhao, Q., Wang, M., and Wu, G. (2018) A new promising nucleating agent for polymer foaming: effects of hollow molecular-sieve particles on polypropylene supercritical CO_2 microcellular foaming. *RSC Advances*, **8**, 20061-20067.

72. Zhu, H., Monrabal, B., Han, C. C., and Wang, D. (2008) Phase structure and crystallization behavior of polypropylene in-reactor alloys: insights from both inter-and intramolecular compositional heterogeneity. *Macromolecules*, **41**, 826-833.

73. Selden, R., Nyström, B., and Långström, R. (2004) UV aging of poly (propylene)/wood-fiber composites. *Polymer Composites*, **25**, 543-553.

74. *Developments in Polymer Stabilisation - 8*, Scott, G. (ed.), Elsevier, USA (1987).

75. Popa, M. I., Pernevan, S., Sirghie, C., Spiridon, I., Chambre, D., Copolovici, D. M., and Popa, N. (2013) Mechanical properties and weathering behavior of polypropylene-hemp shives composites. *Journal of Chemistry*, **2013**, Article ID 343068.

76. Onggo, H., and Pujiastuti, S. (2010) Effect of weathering on functional group and mechanical properties of polypropylene-kenaf composites. *Indonesian Journal of Materials Science*, **11**, 123-128.

77. Grigoriadou, I., Paraskevopoulos, K., Chrissafis, K., Pavlidou, E., Stamkopoulos, T.-G., and Bikiaris, D. (2011) Effect of different nanoparticles on HDPE UV stability. *Polymer Degradation and Stability*, **96**, 151-163.

78. Al-Shabanat, M. (2011) Study of the effect of weathering in natural environment on polypropylene and its composites: Morphological and mechanical properties. *International Journal of Chemistry*, **3**, 129.

79. Rajakumar, K., Sarasvathy, V., Chelvan, A. T., Chitra, R., and Vijayakumar, C. (2009) Natural weathering studies of polypropylene. *Journal of Polymers and the Environment*, **17**, 191.

80. Hoekstra, H., Spoormaker, J., and Breen, J. (1997) Mechanical and morphological properties of stabilized and non-stabilized HDPE films versus exposure time. *Macromolecular Materials and Engineering*, **247**, 91-110.

81. Rabello, M., and White, J. (1997) Crystallization and melting behaviour of photodegraded polypropylene-I. Chemi-crystallization. *Pol-*

    ymer, **38**, 6379-6387.
82. Rabello, M., and White, J. (1997) Crystallization and melting behaviour of photodegraded polypropylene-II. Re-crystallization of degraded molecules. *Polymer,* **38**, 6389-6399.
83. Craig, I., White, J., and Kin, P.C. (2005) Crystallization and chemicrystallization of recycled photo-degraded polypropylene. *Polymer,* **46**, 505-512.
84. Schoolenberg, G., and Meijer, H. (1991) Ultra-violet degradation of polypropylene: 2. Residual strength and failure mode in relation to the degraded surface layer. *Polymer,* **32**, 438-444.

Index

■ Index